THE

BENZO

BOOK

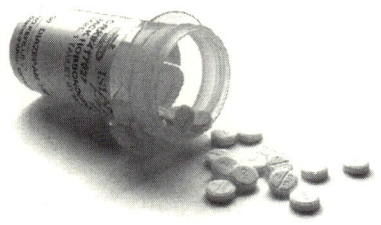

getting safely off tranquilizers

Jack Hobson-Dupont

ESSEX PRESS NANTUCKET

ALL TRADEMARKS, REGISTERED TRADEMARKS, COPYRIGHTS AND SERVICE MARKS APPEARING HEREIN ARE THE PROPERTIES OF THEIR RESPECTIVE OWNERS AND ARE HEREBY ACKNOWLEDGED.

1ST EDITION ✹ ESSEX PRESS

COPYRIGHT © 2006, JACK HOBSON-DUPONT
ALL RIGHTS RESERVED :: PRINTED IN U.S.A.
ISBN 978-1-4116-9259-6

Dedicated to Janie, and to my heroes: Ray, Yvonne, Julie, James, Deborah, Mary, Jackie, Georgia, Maurice, Carolina, Bob, Layne, Marsha, Jule, Karissa, Steve, Al, Neta, Dan, Pauline, Tracey, Gretchen, Ayla, Koren, Glenn, Chris, Karen, Geraldine, Susie, Pip, Elsa, Lisa, Jesse, Ron, and so many, many others....

Contents

CHAPTER ONE: In Tolerance 1

CHAPTER TWO: Xanax 8

CHAPTER THREE: The Ashton Method 12

CHAPTER FOUR: Ray Nimmo's *benzo.org.uk* 16

CHAPTER FIVE: My First Idea 21

CHAPTER SIX: Contradictory Beliefs 24

CHAPTER SEVEN: Down-Regulation 34

CHAPTER EIGHT: Drugs and the FDA 49

CHAPTER NINE: Withdrawal Symptoms 71

CHAPTER TEN: Supplements 85

CHAPTER ELEVEN: The Emotions of Survival 92

CHAPTER TWELVE: *"Don't take it personally"* 97

CHAPTER THIRTEEN: Persistent Depression 105

CHAPTER FOURTEEN: The Chemical Imbalance Theory . 108

CHAPTER FIFTEEN: Market-Driven Medicine 118

CHAPTER SIXTEEN: Haywire 125

CHAPTER SEVENTEEN: Thinking and Feeling 132

CHAPTER EIGHTEEN: Therapy During Recovery 139

CHAPTER NINETEEN: Compassion Burnout 143

CHAPTER TWENTY: The Lower Doses 148

CHAPTER TWENTY-ONE: Fear 155

CHAPTER TWENTY-TWO: The Last Dose 163

CHAPTER TWENTY-THREE: *"Ask your doctor"* 170

CHAPTER TWENTY-FOUR: Addiction *vs.* Dependence . . 186

CHAPTER TWENTY-FIVE: Getting Safely Off Benzos . . 198

CHAPTER TWENTY-SIX: Post-Benzo 206

APPENDICES

INDEX

*The signpost on the road to Hell
reads: Heaven, straight ahead*

Chapter One
In Tolerance

Perhaps because of decreased atmospheric pressure of the air, there is a quality to the light right before a hurricane comes that is exceptional. I came down from clouds suffused in that preternatural light and landed at Norman Manley International Airport in Kingston, Jamaica. I went straight to my hotel, arriving just as the first bands of gray, rain-laden clouds began to appear. I remember feeling a perverse thrill at the thought that I had knowingly gone to a place even as a hurricane bore down upon it. But the trip was important. My brother and I had a small company together and had, with considerable difficulty, negotiated a deal to sell high quality waste removal trucks to the Jamaican government. On the strength of this first deal, we then hoped to provide the country with a state-of-the-art composting system for solid waste management. While not a very glamorous adventure, I was rewarded by the thought that a modernized facility for processing garbage would cut down on the associated disease vectors, and that because of my efforts, people would lead healthier lives.

The deal was essentially done but there was one final piece to be concluded and that was why I was in Jamaica. I had to meet with an attorney associated with a certain engineering company whose owner was a powerful force in the local political environment. As such, my job was to wait at my hotel until the attorney called to arrange a meeting.

Perched high in a luxury hotel I watched, fascinated, as the hurricane struck. The shrieking wind, the rain rattling against the windows, lights flickering as the electrical power fluctuated, I found it all exciting. Compared with the hurricanes that would, years later, devastate much of the Gulf Coast of the United States this storm was quite a minor affair but its raw power and beauty were thrilling to behold. The hurricane passed over the island and in the days that followed it schedules were understandably in disarray. There was nothing for me to do but to wait for the phone call from the attorney.

I would awaken in the morning, shower, and lay out my business suit on the bed so it would be ready for me to put on at a moment's notice. Then I would sit around half-dressed in the moist, tropical air reading books I had brought, and playing my electric guitar through headphones. I am usually very content to have the luxury of spending time by myself but during those days of waiting to meet the attorney I noticed that I began to feel ... peculiar. It was a vague sort of sensation, making me feel slightly fluttery and disoriented, and compelled to rest, though I was certainly not tired. Something was definitely wrong with me, but it didn't seem like I was sick in any way. The days began to melt into one another, and sleep became strange.

One day as I tried to puzzle out what I was experiencing I had an intuition that I needed chocolate, of all things. I got dressed, went down to the shop in the lobby and bought a chocolate bar. I brought it back to my hotel room, then proceeded to break off careful pieces and eat them. After three small pieces, I felt the tiniest amount more like myself again. I set the chocolate bar aside and would, over the next three days, take pieces of it as though medicinally.

I never did meet the attorney. Rather, I met with his sister, herself a powerful lawyer. As she wouldn't tell me anything directly, I learned from an associate that the deal had fallen through at the last minute. A political intrigue amongst the players had caused a rift between self-interested factions. I flew back to the States. Later I learned that the government had bought inferior trucks, and even later, I heard that the trucks were rusting in the municipal yards.

After my return from Jamaica the 'peculiar' feeling engulfed any ability to forestall it with pieces of chocolate. Over the next month my condition deteriorated. A profound fatigue overcame me, fatigue so great it seemed impossible to raise my arms at times. I was constantly exhausted yet at the same time unable to sleep. I felt so nervous that I was unable to tolerate any stimulation at all. I couldn't listen to music, couldn't watch television. The only comfort I seemed to have was reading stilted mystery novels from the 1930's right before going to bed.

A brutally cold winter settled in and my condition worsened. The cottage I live in was heated only by a wood stove and under normal conditions, that's enough for me. I would find it bracing to awaken to a cold house, then have to cut kindling and light the fire. I was now so weak, however, that even to move at all was daunting. It seemed like torture to light that fire in the fireplace. Moreover, the cord of wood I had bought that autumn must have been improperly seasoned because it wouldn't stay lit without constant poking. By the end of the next month, I was reduced to a state in which I simply sat on a futon on the floor, day and night, stoking my fire and trying to survive.

I would curl up on the futon to attempt sleep, but sleep was sporadic at best. My mental state deteriorated. I became unable to tolerate stimulation. The ringing of the telephone was like someone banging a cooking pot next to my ear, my nerves jumping at the impact of the sound. I turned off the telephone ringer to spare myself from it. I wore a sweatsuit, day in, day out, and simply continued to sit on the futon. The most difficult thing in my life became the shower. As each day passed I would become increasingly grimy, and feel desperately the need for a shower. It was hard to drag myself into the bathroom and turn on the taps. And then, to take off my sweat suit in the chilly air and stand in the stream of water was far, far too stimulating an experience for me to endure. Many times I would make it as far as starting the shower, only to fail to rouse the energy necessary to disrobe and get into the tub. I'd turn off the taps and retreat to my futon on the floor, hoping for more success the following day. A week, two weeks, sometimes three weeks would pass before I could summon the strength, both physically and mentally, to take that shower, and when I did, the sensation of standing in the jet of water, being alternately cool on one side of my body and then the other, was excruciating, almost unbearable. Toweling off after the shower was itself a challenge. I could barely handle the sensory stimulation, and would tug my sweatsuit back on gratefully, then collapse on my futon under a blanket.

What on earth was wrong with me?

During the previous summer, I had contracted babesiosis from a tick bite. A course of treatment had rid me of the parasites, but I had never seemed to recover. In my most capable moments, I would seat myself before my computer and search the Internet for clues to what was happening to me. My symptoms seemed to be those of acute Chronic Fatigue Syndrome, and I found that there is a variation of that condition known as Post Viral Fatigue Syndrome. Like Chronic Fatigue Syndrome this was not a condition whose diagnosis could be arrived at scientifically. In other words, there were no medical tests I could undergo whose results would indicate the presence of a virus, or the absence of anything necessary for health. It was once thought that Epstein-Barr Virus (EBV) caused Chronic Fatigue Syndrome, but that idea had been discredited. Many people who have never had a related sickness show antibodies against Epstein-Barr Virus in blood tests, indicating they had at some time been exposed to EBV. At most it can be said that Epstein-Barr is 'associated' with Chronic Fatigue Syndrome and mononucleosis, but not that it causes those conditions.

So, I labored under the idea that I had a fatigue syndrome as a result of my having had babesiosis and I despaired of ever getting much better. After a number of months, a thought struck me. I began to wonder if my condition weren't somehow psychosomatic. After all, aside from the physical debility, I was most definitely mentally impaired as well. My thinking was foggy, unclear, and my emotional state was dismal. I had formerly taken an antidepressant, having weaned myself off of it in the previous year. What if my condition were nothing more than some form of depression? What if my body had grown dependent upon the antidepressant and was malfunctioning now because I no longer took it?

As an experiment, I decided to take a small amount of the SSRI-type antidepressant I had been on, Effexor. I broke one of the 'jagged little pills' and took half of it, an amount that was twenty-five per cent of my previously prescribed dose. All hell broke loose. When that small amount of Effexor hit my central nervous system, a grinding sensation ran through my entire body. My thoughts raced out of control. Whenever I closed my eyes, I was plunged into a roiling mental chaos, like viewing three dreams at once while listening to a dozen radio stations at the same time. It would disappear instantly when my eyes opened, return when I again closed them. I began to tremble uncontrollably, at times violently. An unspeakable horror filled me and didn't let up. Lying on my futon in a state of utter agony, both physical and mental, I would look up at the clock and see that only five minutes had passed since the last time I looked, although it had seemed interminable. Hour after hour, day after day, passed at the slow cadence of five-minute segments of abject horror. I lost twelve pounds in eight days.

I lost my mind. To be more accurate, I lost the access to a large part of my mind. I could think analytically, but, oddly, that was about all. I lost the ability to think creatively or to have ideas. And bizarrely, I could think of absolutely nothing other than myself and the condition I was in. Even after the effect of taking the drug wore off, I was still unable to think of anything other than 'me.' I was unable to consider my son, my wife, my brother or my many friends. By this I mean *every* thought was about myself. If someone was speaking to me about something they felt or experienced, I could only relate it to myself and my own experience. I was pathologically self-centered to the absolute exclusion of everything else.

After a little more than a week I had improved, but only to the extent that the more dramatic effects had diminished somewhat. The only relief I had was for a few minutes at night, when I would read before closing my eyes to go to sleep. It suddenly struck me that, by long habit, I would take my Xanax tablets before going to bed. Perhaps the calming effect wasn't from reading, but from the Xanax? I thought, "I doubt that those little white pills can have much of an effect on me, but I'd better be thorough and see if there's anything in the medical literature about Xanax."

I had been prescribed Effexor almost ten years previously at the time when Prozac and other Selective Serotonin Reuptake Inhibitor (SSRI) type antidepressants were commonly, almost routinely, being offered. I found it to be wonderful. After five days of being on Effexor, I bounded out of bed with the energy and optimism of an eighteen year-old, and every day thereafter was a similarly enthusiastic experience. The only problems with it were that I was slightly nauseous for a few hours after taking it and I felt strangely and uncomfortably stimulated by it. When I told these symptoms to my doctor, he prescribed Xanax to counter the 'agitation' I had reported. I began taking the Effexor right before going to sleep so that the nausea would occur while I was asleep, and I took the Xanax at the same time. The problems went away, and I got on with my life with the vigor the antidepressant gave me.

I had been a little concerned when my doctor suggested the Xanax because I knew it was a tranquilizer and I was worried that it might be addictive. I had had substance abuse problems decades earlier in my twenties, and therefore, felt myself to be at risk. My doctor assured me that it was 'safe' so, I never gave it another thought. While taking these drugs year after year, I surmised that I must certainly have developed, not an addiction, but a mild physical dependency upon the Xanax. I continued taking Xanax even after I had stopped taking Effexor because of that presumed dependency and had figured that, when I could take the time off from my business, I would need to check into a detox clinic to have my Xanax problem 'taken care of' professionally.

Chapter Two
Xanax

Wondering how the Xanax might be affecting me, I googled 'Xanax' on the Internet. Among countless offers for buying Xanax without a prescription online, there were a number of sites offering authoritative medical information, describing the uses of *alprazolam* (the generic name of Xanax) and its dosage, side effects, interactions with other drugs, and other such things. A thorough reading indicated nothing that seemed to imply that there was anything problematic about the drug, nor anything to suggest that it might be involved in my deplorable physical and mental condition.

Further down in the internet search results, however, I found references to bulletin board type websites where individuals could share their own experiences. Suddenly, I was immersed in descriptions of the same sort of horrors I had been going through. To every cry for help the response was the same: the drug was the cause of the problem. "That's impossible!" I thought. But I read on.

Those who had gone through withdrawal from Xanax or drugs of the same family and were eager to help others urged people to read the work of Professor C. Heather Ashton, a British psychopharmacologist and the leading expert in the world on the subject of tranquilizers. Links were provided to a book she had written, *The Ashton Manual*, which was available online. I followed one such link.

I read *The Ashton Manual* and was appalled. There it was, the explanation of what was happening to me. Evidently, my years of Xanax use had led to a chemical dependency on the drug. Since I had not kept steadily increasing my dosage of it, I had been experiencing what is known as *interdose withdrawals*. Xanax is a fast-acting drug with a short 'half-life', *i.e.,* the amount of time it takes for half the metabolites of the drug to leave the system. By not ramping up the dosage to match the tolerance my body had developed for Xanax, I was apparently going into withdrawal every day as the dose taken the previous night wore off.

The actual title of what is commonly referred to as *The Ashton Manual* is *Benzodiazepines: How They Work and How To Withdraw*. I learned that Xanax is one of a class of drugs called *benzodiazepines* which share a common chemical basis. I was a bit stunned to realize that so many drugs, which are presented as being quite different from one another, are based upon the same chemical compound. Thus it is that a sleeping pill such as Restoril or Dalmane is chemically related to tranquilizers like Xanax or Halcion, and even to a substance such as Rohypnol, which is described in news articles as 'the date rape drug.' Even more shocking to me was the realization that these drugs had been derived from the same underlying chemistry as Valium.

I was very much a child of 'The Counter Culture' of the late 1960's and early 1970's. I had been a musician during this time, and recreational use of drugs was common among my peers, both organic drugs such as marijuana, as well as pharmaceutical drugs, *e.g.,* codeine, tranquilizers, sleeping pills, and the like. During those years, I had abused such pharmaceutical drugs and I had, on many occasions, regrettably, popped Valium tablets as part of an evening's entertainment. As my twenties—and the chapter in my life where I used drugs—were drawing to a close, I recall hearing from my peers warnings about the use of Valium to 'get high.' Valium, it was being reported, was apparently highly addictive, far more difficult to get off of than heroin. And unlike heroin, Valium users who attempted to quit 'cold turkey' faced the possibility of going into convulsions and dying. After hearing such rumors, I steered clear of Valium, refusing it when it was offered to me.

Just after I had emerged from this phase of drug abuse, I saw a movie starring Jill Clayburgh called, *I'm Dancing As Fast As I Can,* from the book of the same name by Barbara Gordon. It was a terrifying look into the life of a professional woman who struggled to overcome an addiction to Valium. Watching that frightening movie, I remember being grateful that I had been spared such an ordeal. Around the time of that movie, there was recognition that vast numbers of people in the United States, Canada and the United Kingdom, mostly women, were addicted to Valium, and efforts were being made to wean these people off the drug.

So, imagine my shock at reading *The Ashton Manual* and learning that the Xanax I had been prescribed over a period of almost a decade was essentially the same drug as Valium. And there was worse news. Professor Ashton had developed an equivalency table to compare the relative potency of the various benzodiazepine drugs. From it I learned that the Xanax I had been taking was *twenty times stronger than Valium,* In other words, one milligram of Xanax was the equivalent of taking *twenty milligrams of Valium.*

I felt absolutely betrayed. When my doctor suggested that I take Xanax to combat the agitation that accompanied my antidepressant use, I had specifically asked if it were dangerous. Since I had a history of drug use decades earlier, I was especially careful to avoid anything that would put me at risk for addiction. My doctor assured me that Xanax was 'safe and effective.' If I had had the slightest idea that Xanax was based upon the same chemical compound as Valium, but twenty times more potent, I would certainly never have taken it. I would either have learned to tolerate the agitation I was experiencing with Effexor, or gone off the Effexor altogether.

A basic tenet of ethical medical practice is that of 'informed consent.' A doctor presents a patient with the facts concerning a treatment or procedure, spelling out the various dangers as well as the possible benefits, and the patient then makes medical decisions based upon that information. Not to disclose the full risks of substances such as benzodiazepine drugs promotes not merely uninformed consent but *mis*-informed consent by patients. As a result of misinformation, I now found myself in the worst possible medical trouble, severely physically and mentally compromised; and a long way from being well.

Chapter Three
The Ashton Method

From *Benzodiazepines: How They Work and How To Withdraw* I learned that the method Professor Ashton had developed in her years operating a clinic to help people wean themselves off of benzodiazepine-based drugs was to transition the patients from whatever drug they were using onto Valium, then to reduce slowly their daily dosage of Valium. The chief virtue of Valium is its long half-life. It takes about eight days after a given dose for its metabolites to fall by fifty percent, signifying that the blood concentration of Valium (or, *diazepam,* its generic name) remains consistent over a considerable period of time, contributing to a smoother experience of the drug's effects than those of faster-acting benzodiazepines. I was determined to get myself quit of Xanax and this method appeared to be the best—which is not surprising considering that the

Ashton Method is the only benzodiazepine discontinuation protocol that is based upon both rigorous scientific research as well as actual clinical experience.

The Ashton Manual contains dosage equivalency charts comparing diazepam to related drugs, and 'schedules,' *i.e.,* tables to show given dosages of diazepam reduced over time. Using these resources, I was able to devise a tapering schedule for myself. Since I had been taking 3 mg of Xanax, and since 1 mg of Xanax is the equivalent of 20 mg of diazepam, that meant that my initial dosage of Valium needed to be 60 mg of diazepam daily. *Sixty milligrams of Valium.* I could hardly believe I'd have to take such a massive quantity of Valium to approximate the amount of Xanax I was on—but there it was in *The Ashton Manual,* and if anyone would know the correct equivalencies of these drugs, it would be Professor C. Heather Ashton, DM, FRCP.

Professor Ashton is the author of more than fifty published papers about benzodiazepine drugs. Holding multiple degrees in Medicine from the University of Oxford, she became a Member of the Royal College of Physicians in 1958, and a Fellow of the Royal College of Physicians in 1975. She has served as National Health Service Consultant in both Clinical Psychopharmacology and Psychiatry.

Now Emeritus Professor of Clinical Psychopharmacology at the University of Newcastle upon Tyne, in Britain, her research at the university focused on the nature of psychotropic drugs and their effects upon the brain and upon behavior. She operated a benzodiazepine withdrawal clinic for twelve years, and has provided expert testimony about benzodiazepines in both governmental investigations as well as litigations. She continues to lecture and serve as a consultant.

Therefore, as its author is arguably the world's foremost expert in these drugs, if *Benzodiazepines: How They Work and How To Withdraw* stated that a dosage of 60 mg of diazepam is the correct equivalent of

3 mg of Xanax, I felt reasonably confident that that was accurate.

I made an appointment to see my doctor, and, armed with a copy of *The Ashton Manual* and a printout of the withdrawal schedule I had come up with, I told him that I wanted to get off of the Xanax and wanted to use the Ashton Method in order to do so. I had no idea if he would accede to my request. I was angry with the doctor for having facilitated my becoming an unwitting drug addict—but I was far more concerned with rescuing myself from the addiction than venting my anger, and I would need his help. He read through the materials and considered the proposal carefully. He then told me he thought it was a novel approach to getting off of tranquilizers but that if this was what I had chosen to do, he would do his part to implement the method and to monitor my progress with it. I felt a great sense of relief, and left with a prescription for diazepam.

Presenting that prescription at the pharmacy, however, produced wrinkled brows and expressions of suspicion in the pharmacists. I showed them the protocol I was using, but they were reluctant to prescribe a drug like Valium, especially in such large amounts. The pharmacist called my doctor and consulted with him over the telephone before agreeing to fill my prescription. When I finally made it home, I collapsed on my futon couch. I had exerted more energy that day than I had in months. I had been incapable of much more than sitting in front of the fire, but this was a matter of survival so some resource deep within me kicked in and gave me the ability to exert an extreme effort to go out and accomplish the procurement of diazepam tablets to begin the lengthy process of getting myself off of Xanax.

The first stage of the Ashton Method calls for a gradual, step-wise shift to diazepam from whatever benzodiazepine drug one is using. So, my first dose that night was to consist of 2 mg of Xanax with 20 mg of diazepam. I took my pills and went to bed—where I slept for a total of two hours. Because of the slower action of diazepam, there was simply not enough benzodiazepine in my system to sustain sleep longer than two hours that first night. In the days that followed, this improved; but I never slept more than about three or four hours per night over the next two years or more. The important thing, though, was that the process had begun.

Chapter Four
Ray Nimmo's *benzo.org.uk*

During the period when I crossed over from Xanax to diazepam, I began to feel more and more sedated. Since diazepam has a longer half-life, the metabolites of the drug are active in the system for longer periods of time; so, while Xanax would clobber me with its full effect, causing me to fall asleep, then abruptly leave my system four to six hours later, the Valium would exert its influence more noticeably throughout the day. It was ironic that after almost a decade of using benzodiazepine, the first time I felt 'high' from it was when I was beginning the process of weaning myself off of it.

And I didn't like the sensation of feeling intoxicated that it gave me. When I had abused drugs in my twenties I enjoyed feeling high, but came instead to cherish above all the feeling of mental clarity. I became quite protective of that clarity. As a result, I never drank alcohol—I simply didn't enjoy clouding my mental processes. Occasionally, I would have a few glasses of champagne at a wedding and when the alcohol would hit me, I'd remember only too clearly why I didn't drink. And now, here I was, oversedated as a result of taking Valium.

I had a tendency to slur my speech and was often unable to find the right word. That was particularly irritating, as I'd always had a fairly expansive vocabulary. There was, however, a certain relief that the sedation imparted and because I had been in such a horrible mental state the relief was welcome. My mind was still far from functioning properly, though. I still could not form actual ideas—and words cannot express how disturbing that was. My thoughts were quite rational, thankfully, but they centered only on myself and the condition I was in. Thoughts of a creative nature were simply unavailable to me. Even an 'idea' as simple as, "if I have a snack now, I won't feel like eating dinner later" was an impossibility.

That was devastating to me. I had designed and programmed computer software and been a troubleshooter of computer systems for most of my professional life. Solving problems by thinking up then implementing ideas was the essence of how my mind operated, and now those abilities were absolutely gone. I had been a musician and a painter as well, and not only was my talent for ideation gone, I had absolutely no æsthetic sense. I would look at the paintings on my walls and they had no more emotional impact upon me than a traffic sign would have done. Worse still, when I tried to play my guitar, the effect wasn't merely neutral like my paintings; the sound of the music would grate upon my nerves like fingernails on a blackboard.

I had already spent a considerable amount of time unable to read, listen to the radio, or watch television. Getting from one moment to the next, then the next, then the next, felt like I was 'hanging on for dear life.' After I began tapering with Valium, I found that I became capable of watching movies on television. I still couldn't watch the news or regular TV shows—they were far too stimulating, the audio and visuals seeming to be designed to crank up the nervous system—but movies were a blessing. Trying to follow the plot gave me something for my poor mind to latch onto for an hour and a half. I could never actually lose myself in it, could never really forget how wretched I was feeling, but movies did provide some distraction, and watching movies in the evening—even bad movies—gave me something to look forward to all day.

The emotions I was capable of having were fear and a pervasive and absolute misery, a sort of perpetual agony I would later hear people refer to as 'benzo hell.' The only thing that sustained me was the knowledge that I was following a protocol that would get me off the wretched drugs that had caused me to be in that awful state. I continued to search the Internet for more information about what had happened to me and I found an on-line repository of basically everything that is known about benzodiazepines, called *'benzo.org.uk'*. A man named Ray Nimmo in Britain hosted it. Ray Nimmo is a remarkable man.

In 1984, after having an allergic reaction to an antibiotic, Ray was told by his doctor that the abdominal pain he was experiencing was a muscle spasm. The doctor prescribed Xanax as a muscle relaxant. When Xanax failed to address the problem, Ray was prescribed Valium and was kept on it for the next fourteen years. His doctor told him that he needed to be on the medication. As Ray put it himself during an interview on the BBC in 2002, he was, ". . . suicidally depressed, so anxious, agoraphobic, lethargic. I just didn't want to go out of the house. I didn't want to answer the door or the telephone. I was just like a zombie."

Unable to work, Ray lived in a "twilight world of paranoia and fear." His doctors told him that he was suffering from a mental illness and he trusted that, as doctors, they knew best. It was only in 1998 that another doctor, a surgeon at a local hospital, examining Ray after an ultrasound treatment, told him that his problems were due to being on diazepam. Another doctor confirmed this view and diagnosed him as having Valium-induced depression as well as Valium addiction. With this doctor's guidance, Ray gradually reduced his intake of drugs over the next three months. The agoraphobia, anxiety and suicidal depression he had suffered with for fourteen years—even the abdominal pain that had caused the original prescription of benzodiazepine—resolved. Ray once again felt like himself.

Responding to the injustice of having had fourteen years of his life ruined, Ray Nimmo sued the physician who had kept him on drugs for all that time. He was awarded £40,000 in 2002 and went on to become a champion for the cause of benzodiazepine overprescription in the United Kingdom. He developed the *benzo.org.uk* website, and initiated an on-line community where fellow victims of benzodiazepine could share their experiences.

I was in pretty bad shape when I discovered Ray's website and joined the discussions there. The counsel and compassion of the volunteer administrators and moderators were of invaluable help, especially during those first days of my utter bewilderment and confusion about what was happening to me. The camaraderie of others who were also now using diazepam to taper off of benzodiazepines proved essential—just to know I was not alone in the ordeal was a great comfort. I was continually amazed at the humanity and compassion these people showed one another, as well as their seemingly endless capacity for providing support, which only someone who had experienced the rigors of withdrawal could understand well enough to give.

I don't know what would have happened to me had I not found Ray Nimmo and his website. I am humbled by the magnitude of the gift that those at *benzo.org.uk* gave to me and countless others. I would spend the next two and a half years checking in at that website on a daily basis, sharing with others the daunting process of saving our own lives from the oblivion to which benzodiazepines had consigned us.

Chapter Five
My First Idea

The brutally cold winter was winding down as I continued the crossover from Xanax to Valium, a process that took one month to complete. On sunny days I would sit outside on the back deck just to break up the monotony of my existence, which usually consisted of sitting in front of the fireplace wrapped in a blanket, stoking the logs ceaselessly, or sitting in my chair at the computer while I scoured the Internet for information that might help me. The fire would go out overnight in the few hours while I slept, so I woke with the urgency of needing warmth to survive. The temperature inside my cottage was usually just a few degrees above freezing in the morning, making the cutting of kindling and the lighting of the fire an automated frenzy, often done with chattering teeth and shaking hands. As such, the days that were warm enough for me to sit bundled up on the back deck soaking in sunlight were a relief from struggling against the gloomy chill of winter.

As I had become less and less functional the previous autumn I simply abandoned everything that had previously comprised my life. I stopped my professional involvement in my business endeavors with my brother, stopped my web design and consulting business, stopped the routine maintenance on my house. I just couldn't cope with any of it, nor had I the energy to do much of anything other than to lie in front of the fire, just trying to endure from one moment to the next. My isolation was almost complete. My son was in his senior year in high school and I avoided him because I didn't want him to see me in such a weakened, devastated state. The only person who saw much of me was my wife. She would cook dinner each night—a comfort I came to treasure—but often I simply didn't have the energy to maintain a conversation, so we ate in silence.

Not having the strength to meet my responsibilities, I simply neglected them. I had bought a vintage Airstream travel trailer and had been restoring it, but that project, too, had been abandoned. The trailer was parked next to the east side of our house, but it was encroaching on our neighbor's property and I knew I'd have to do something about that before he and his family came up for the summer. That thought was a constant aggravation to me. The question was, what could I do?

My own property is very small, the back yard bordered by hedges. One day, sitting in the sun on the back deck, idly staring at the back yard, I thought, "You know, if that pine tree weren't there in the corner, I could back the trailer in along the southern edge of my property and the cedars would hide it from view." And that was the first actual idea I had had in over a month.

For me to look at that pine tree which had been a fixture of the property ever since we had owned our house and visualize it *not* being there—well, that was actual creative thinking and problem-solving. Until that moment, I had not been capable of any such ideation. Rather, my thinking had been a sort of furious rationality as I examined information on the Internet and assessed whether or not it bore upon my condition. It would be another month before I would be able to muster the energy to cut that scraggly old pine tree down and move my Airstream trailer, but the important thing was that I was again capable of having ideas. Slowly, too, as Valium supplanted the Xanax in my body, my æsthetic sensibilities were restored. I didn't have the strength, the will or the enthusiasm to appreciate my paintings again, but at least I could tell what they were. These were the very first signs that I might actually recover from what had happened to me.

Chapter Six
Contradictory Beliefs

All the devastation that had happened to me happened as a result of my having taken Xanax. But how could that be? How could a medicinal product in widespread usage, believed to be 'safe and effective,' have caused me to become not only debilitated by it but addicted to it as well? Xanax is a popular drug. According to the January 1993 issue of Consumer Reports, Xanax was the largest-selling psychiatric drug in the United States, and the fifth most frequently prescribed drug in the country, given to millions of Americans. Obviously, an overwhelming majority of the people who take it must be able to discontinue its use without being hit with extraordinary difficulties or else these difficulties would be widely known throughout the medical professions. And yet, I could hardly be alone in my reaction. So, what percentage of Xanax users does suffer as a result of using it?

Unfortunately, there is no clear answer to that question. According to *Beyond Benzodiazepines,* the manual of TRANX, a non-profit organization established in Australia to educate people about the dangers of benzodiazepines and help them safely discontinue use of the drugs, while *"not all people become dependent upon benzodiazepines,"* and *"some long term users of these drugs stop taking them without ill effects,"* a full fifty to eighty percent of those who have taken benzodiazepines for over six months *"will experience withdrawal symptoms when reducing the dose."* Given the vast numbers of people who use such drugs, that such a high percentage of them will have withdrawal symptoms seems exaggerated. And yet, the assessment of TRANX is based upon their years of experience with actual people struggling with dependence upon benzodiazepines.

In fact, that there even are such organizations as TRANX and others suggests that the problem is more widespread than is commonly known. A policy paper put out by the British Columbia Centre of Excellence for Women's Health notes, *"It is estimated that 3 to 15% of any adult population is using and may be addicted to benzodiazepines. Of this group, 60 to 65% are women."* In the United Kingdom, there are volunteer groups and government initiatives to help reduce the number of people prescribed benzodiazepines over lengthy periods. It is estimated that, in the British Isles, one million of its people take benzodiazepines habitually. Paul A. Quigley writes in *Public Health Dimensions of Benzodiazepine Regulation,* "Despite *well-known problems of dependence and misuse, they continue to be widely prescribed, even on an unlicensed long-term repeat basis, particularly to women, the elderly, the chronically ill and other groups of people who suffer social and educational disadvantage. Guidelines discouraging such prescribing have been relatively ineffective. . . ."*

What is most peculiar about such long-term use is that the medical regulatory agencies in Australia, Canada, the European Union, the United Kingdom and the United States have all issued the guidelines referred to above. They warn doctors that, due to the danger of dependence, benzodiazepines should not be prescribed for a period longer than two to four weeks. In issue No 41, p. 166, of its advisory publication, The British National Formulary (akin to the Food and Drug Administration in the U.S.) states:

> *Benzodiazepines are indicated for the short-term relief (two to four weeks only) of anxiety that is severe, disabling or subjecting the individual to unacceptable distress, occurring alone or in association with insomnia or short-term psychosomatic, organic or psychotic illness.*
>
> *The use of benzodiazepines to treat short-term 'mild' anxiety is inappropriate and unsuitable.*
>
> *Benzodiazepines should be used to treat insomnia only when it is severe, disabling, or subjecting the individual to extreme distress.*

In January of 2004, in a communication from the Chief Medical Officer in the U.K. to all doctors, the position was reaffirmed with a *'Benzodiazepines Warning'* concerning patient safety: *"Doctors are being reminded that benzodiazepines should only be prescribed for short-term treatment, in light of continued reports about problems with long-term use."*

The European Union's *Guidelines*, VOLUME 3B, states it more definitively:

> *Use of benzodiazepines may lead to the development of physical*

and psychic dependence upon these products. The risk of dependence increases with dose and duration of treatment.... The duration of treatment should be as short as possible depending on the indication, but should not exceed 4 weeks for insomnia and eight to twelve weeks for anxiety, including the tapering off process. Extension beyond these periods should not take place without reevaluation of the situation.

In fact, not only does use of benzodiazepines beyond these time periods increase the risk of dependence, there is evidence that the very effectiveness of the drugs to treat patients' problems for more than a short time is dubious. In the U.K., the Committee on the Review of Medicines (CRM) was cited in *The British Medical Journal*, 29 March, 1980:

The committee further noted that there was little convincing evidence that benzodiazepines were efficacious in the treatment of anxiety after four months' continuous treatment. It considered that an appropriate warning regarding long-term efficacy be included in the recommendations, particularly in view of the high proportion of patients receiving repeated prescriptions for extended periods of time.

It further suggested that patients receiving benzodiazepine therapy be carefully selected and monitored and that prescriptions be limited to short-term use.

The British charity, Adverse Psychiatric Reactions Information Link (APRIL), advises:

In the UK many people obtain drugs in this group with repeat prescriptions and few patients are re-assessed or warned about the addictive and damaging side effects. . . . In spite of government directives to doctors in the UK, about 12 years ago, that drugs in this group should not be prescribed for more that one month, we hear from many people who have been taking them for 10 years. Of course there are many people who have been taking these drugs for 30 to 40 years.

Manufacturers now recommend no more than two weeks' supply should be prescribed and in all cases withdrawal should be complete within 4 weeks.

The physical and mental damage to some people is tragic. Yet with more information about the dangers of addiction and warnings when first prescribed sleeping tablets or tranquilizers, patients may have been able to avoid the long term damaging effects.

Some people have been helped in the short term by this medication, while others are disabled by it. The problem of overprescribing and benzo addiction is widespread. It seems only government intervention will prevent continuous over prescribing and misuse of this group of drugs.

In *Benzodiazepines—Side Effects, Abuse Risk and Alternatives,* Dr. Lance P. Longo and Dr. Brian Johnson write:

Tolerance to all of the actions of benzodiazepines can develop, although at variable rates and to different degrees. Tolerance to the hypnotic effects tends to develop rapidly, which may be beneficial in daytime anxiolysis but makes long-term management of insomnia difficult. Patients typically notice relief of insomnia initially, followed by a gradual loss of efficacy. Tolerance to the anxiolytic effect seems to develop more slowly than does tolerance to the hypnotic effects, but there is little evidence to indicate that benzodiazepines retain their efficacy after four to six months of regular use. Benzodiazepine therapy is often continued to suppress withdrawal states, which usually mimic symptoms of anxiety. Dosage escalation often maintains the cycle of tolerance and dependence, and patients may have difficulty discontinuing drug therapy.

Therefore, patients are put at risk for benzodiazepine dependence even though benefits from the drugs are negligible. Regarding dependence, in its own product literature, Upjohn, the manufacturer of Xanax, states:

XANAX has the potential to cause severe emotional and physical dependence in some patients and these patients may find it exceedingly difficult to terminate treatment.... The following adverse events have been reported in association with the use of XANAX: seizures, hallucinations, depersonalization, taste alterations, diplopia, elevated bilirubin, elevated hepatic enzymes, and jaundice.

So, while short-term use only is clearly indicated, physicians apparently routinely continue to prescribe benzodiazepines for long-term use, in direct contradiction of the guidelines of the various regulatory boards of the nations in which they practice medicine. Even experts in the field disagree with the safety recommendations of national agencies. In *International Study of Expert Judgment on Therapeutic Use of Benzodiazepines,* in the December 1999 issue of the *Journal of Clinical Psychopharmacology,* Dr. E.H. Uhlenhuth, of the University of New Mexico School of Medicine, reports:

Despite decades of relevant basic and clinical research, active debate continues about the appropriate extent and duration of benzodiazepine use in the treatment of anxiety and related disorders. The primary basis of the controversy seems to be concern among clinicians, regulators, and the public about the dependence potential and the abuse liability of benzodiazepines. . . . Overall, the expert panel judged that benzodiazepines pose a higher risk of dependence and abuse than most potential substitutes but a lower risk than older sedatives and recognized drugs of abuse. There was little consensus about the relative risk of dependence and abuse among the benzodiazepines. . . . There was little agreement about the most important factors contributing to withdrawal symptoms and failure to discontinue benzodiazepines. . . . The experts' judgment seems to support the widespread use of benzodiazepines for the treatment of bona fide anxiety disorders, even over long periods.

If 'experts' cannot agree with the conclusions of medical regulatory agencies, we could hardly expect prescribing physicians to do so.

Here we end up with conflicting views—doctors who believe that benzos should only be used in the short-term because of their ineffective performance in long-term use and the danger of dependence, and doctors who believe that extended use is both safe as well as efficacious in treating various disorders. Which of these utterly opposite opinions is true? Are benzodiazepine drugs safe? Or do they pose a danger of harming patients?

We can only conclude that the Food and Drug Administration in the United States, the British National Formulary in the United Kingdom, Health Canada, and the Therapeutic Goods Administration in Australia arrived at the idea of prescribing benzodiazepines for no longer than two to four weeks as a result either of academic studies about the dangers of dependence or as a response to adverse drug reports from a significant quantity of patients. If this be the case, there is something disconcertingly wrong about doctors prescribing these drugs well outside the guidelines the regulatory agencies have established. To do so would mean that doctors dismiss the academic studies and/or the adverse drug reports that resulted in the guidelines having been established.

And yet, there's an even deeper issue that this dichotomous rift in prescribing benzodiazepines brings to light. We are urged at every turn to trust our doctors. We have been taught that physicians all swear to the Hippocratic Oath, the most important tenet of which is, PRIMO NON NOCERE, Latin for: 'First, do no harm.' We believe that doctors won't do anything harmful to us, even while we are suspicious of car salesmen, weathermen, and lawyers. Television commercials for medicines press the phrase, "Ask your doctor if this medication is right for you!" so often that 'ask your doctor' has the brainwashing effect of a 'talking point.'

But if we ask our doctor about the safety of taking a benzodiazepine, we may be told that they are safe only for a period of no more than two to four weeks at which time the dose must be tapered, or we might be told that they are safe for an extended period, and offered an open-ended prescription. The answer we get will depend entirely upon the subjective opinion of the particular doctor.

We trust the medical profession to be run along scientific lines, but this is hardly scientific. In science, there is but one answer to any given question, an answer that has been arrived at by careful and intellectually honest experimentation and observation. Science is tolerant only of objective proofs, not of subjective opinions. Since there are two opposite and conflicting answers to the question of whether benzodiazepines are safe, that undermines the idea that the practice of medicine is based upon sound science.

In earlier times, physicians would drain blood from their patients in the mistaken belief that the body contained four 'humors' which required balancing in order to restore health. Removing quantities of blood would relieve the patient of an excess of one such humor. George Washington, afflicted with a throat infection, was subjected to bloodletting to treat it—and died within twenty-four hours. When medical knowledge expanded to the point where the idea of balancing the four humors was eclipsed, the practice was abandoned. This is exemplary of the progress of any thought-system based in science. It is, however, entirely dependent upon adherents to such a system being willing to learn new ideas, and accept that ideas they may previously have held are false. Doctors, being human, are as prone to clinging to outmoded ideas as the rest of us.

An example of this is the treatment of ulcers. Until very recently, it was believed that ulcers were caused by stress, excess stomach acid, fried foods or other agencies. Treatment was based upon that idea, and ranged from antacid tablets to surgery. In 1982, two Australian research scientists, Dr. Robin Warren and Dr. Barry Marshall, observed a bacterium in the stomach lining of subjects with gastritis. They identified the bacterium as *helicopter pylori* and determined through scientific method that H. pylori was the primary cause of stomach ulcers and duodenal ulcers. (One phase of the scientific method consisted of Dr. Marshall ingesting a small quantity of H. pylori and observing what effect it had upon his gastrointestinal tract.) Having determined that H. pylori was responsible for ulcers, the condition could then be treated with a course of antibiotic treatment.

The new treatment protocol—and the science upon which it was based—was dutifully disseminated throughout the medical world. But it took twelve years for antibiotic treatment of ulcers to be universally adopted and integrated into medical practice. During that time, numerous patients were treated with traditional methods—dietary restrictions, antacids and surgical removal of part of the stomach—rather than antibiotics. The persistence of belief, evidently, caused many doctors to resist the idea that ulcers were caused by a bacterium and not only to hold to their previously-held concepts but to do so at the expense of their patients. It is not only hard but perhaps foolish and irresponsible to maintain implicit trust in doctors, given that they are as prone to such errors in judgment as anyone else.

A parallel may be drawn from this example to the divergent beliefs about benzodiazepines, and provides the only explanation of how, in the population of physicians, there can be two conflicting ideas about the safety and efficacy of the drugs. For me, unfortunately, the whole subject is academic. I asked my doctor when he first prescribed Xanax for me, and he told me they were safe. I have no doubt that he fully believed that that was the case, but what happened to me as a result of taking Xanax was proof that he was horribly wrong about that—as are other doctors who prescribe benzodiazepines for long periods. The most rational assessment would be that such drugs may well be safe for some patients, perhaps even the majority, who are able to discontinue their use without difficulty—but they are not safe for others, for whom benzodiazepines may cause remarkable harm. Anybody taking these drugs would do well to ask himself or herself, *to which group do I belong? The one that's safe? Or the one that is at risk of grievous harm?*

Chapter Seven
Down-Regulation

After having crossed over to Valium from Xanax, I was reducing the dosage by 5 mg per week. I had started on Valium at a 60 mg dose per day, then went to 55 mg, then 50 mg, then 45 mg. . . . I was a bit groggy at these high doses of Valium. The difference in half-life between the Xanax and the Valium formulations explained it. With Xanax, the full strength of the benzodiazepine would hit me, essentially knocking me out so that I slept in a practically comatose state. The Xanax would abruptly wear off while I slept, with the effect that I would wake up in a rather energized condition and dive into the day's responsibilities. Although each ten milligrams of Valium had the same dynamic impact upon me as one milligram of Xanax, the effect didn't dissipate quickly, so I felt it during my waking hours. After the horrors of interdose withdrawal while I was taking Xanax, the sedation of the Valium brought with it a feeling of relief, albeit an unnatural chemical relief at best.

And it was short-lived. As the daily dose of Valium came down week after week, the sedation effect was diminishing. My daily visits to *www.benzo.org.uk* were my salvation. I pestered the administrators and moderators of the website with questions and concerns about what was happening to me, and was provided with insights that could only be offered by someone who had gone through a similar experience. The sense of community among the members of the website was deeply comforting. The utter destruction of what benzodiazepine had done to most of us had reduced us all to a raw state where the luxury of being frivolous or insincere was unthinkable.

I participated in discussions but during those first weeks I mainly drank in the experiences of others, learning the strange vocabulary of BenzoSpeak. Withdrawal Symptoms were known as 'w/d's,' reductions in dosage were called 'cuts' and the effect of cutting was described in detail—different for each person. Helpful advice was mixed with simple compassionate support, shared amongst a group of people for whom mere existence was remarkably, profoundly difficult. And also, there were the so-called 'horror stories.' Perhaps as a cathartic purge or to demystify the bewildering and somewhat horrifying experiences that had resulted from people trying to get off of benzodiazepine, many members of *benzo.org.uk* would tell the story of what had happened to them. Only extremely rarely would it be a tale of drug-abusing addiction. More typically their stories began as mine did, with a well-intentioned physician prescribing a drug in order to help someone. The eventual result of that act of medical kindness, however, was often tragic in the extreme. These 'horror stories' were terrifying and yet it was important, I felt, to get the widest possible sense of the reality of benzodiazepine dependency.

Those who suffered the worst fates, it seemed, were the people who had *c/t'd,* BenzoSpeak for having quit the drug cold turkey. This is a dangerous undertaking. A sudden absence of benzodiazepine in the system may precipitate seizures, severe enough to cause death. Those who had survived were plagued with months—even years—of debilitating effects. This appeared to be the case as well with people who had tapered off benzodiazepine but had done so rapidly, over the course of days or weeks as opposed to months or years, either on their own or at a detox facility. Looking for an underlying rationale, it appeared to me that too great reductions in dosage imposed horrendous stress on the system, that stress induced trauma, and the trauma exacerbated the condition of dependency. I considered the accounts of the people at *benzo.org.uk* and concluded that the wisest course for me to follow would be to reduce my own benzodiazepine intake as slowly as possible.

This was not an easy decision for me to take. I was horrified by the idea that I was some form of drug addict, and humiliated at knowing that I was dependent upon benzodiazepine. I wanted it out of my life. It would be more in keeping with my personality to lock myself in a room, suffer the agonies of withdrawal, then emerge, shaken, but free of the addiction and ready to recover my life. And perhaps with any other addiction that would have been a reasonable solution. But as I listened to what countless others had to say, I realized that a dependency upon benzodiazepine was vastly different than that of any other drug and couldn't be successfully dealt with by means that were appropriate to the normal models of addiction.

Thank heaven for the Internet. Without the ability to connect myself with the ideas and experiences of other people, I never would have figured out what was wrong with me, nor would I have found a method for addressing it. In ignorance that all my troubles were caused by a dependence on Xanax, I may well have ended up in a mental hospital. I would have been given other drugs which might well have obscured forever that benzodiazepine was the problem. Who knows what could have happened to me? And who knows how many people this has happened to, and is happening to right now, but who don't know the role of tranquilizers in their fate?

Reading about the experiences of others with benzodiazepine, I determined that my own goal was not to get off of it *quickly* but to get off of it as *safely* as possible. And for me, that meant to reduce my dosage of it slowly and carefully. The exception to this would be for those individuals for whom the drug itself has a toxic effect. In such a case, a delicate balance would have to be found in order to eliminate the drug from their systems as fast as could safely be done without inducing debilitating withdrawal symptoms. But I seemed to tolerate both Xanax and Valium, so a slow reduction seemed the reasonable course to follow.

That is not to say, however, that the prospect of taking what had been for me such a vile and destructive substance, now that I knew how harmful it could be, was at all appealing. I had made a schedule of dosage reduction, to make it easier to follow the protocol. I saw from the schedule that the amount of time I would be continuing to take Valium stretched out years into my projected future. That idea was overwhelmingly discouraging. But I couldn't figure any other way out of the dilemma. If I reduced more quickly, I put myself at risk for experiencing debilitating withdrawal effects for an undeterminable length of time, well after the last dose of Valium.

To anyone involved in the treatment of addiction, that last statement would seem absurd. How could anything affect someone if it were no longer present in the body? The very idea is illogical. Almost by definition, 'intoxication' is the result of the presence of a substance that is toxic to the body. Remove the toxic substance and the intoxication must necessarily subside. All that could conceivably remain would be psychological effects, such as cravings for the substance.

That is perhaps true of every other substance of habituation—from alcohol to cocaine to heroin to nicotine—but not true in the case of benzodiazepine. Why? The answer requires a bit of understanding about the mechanism by which benzodiazepine acts.

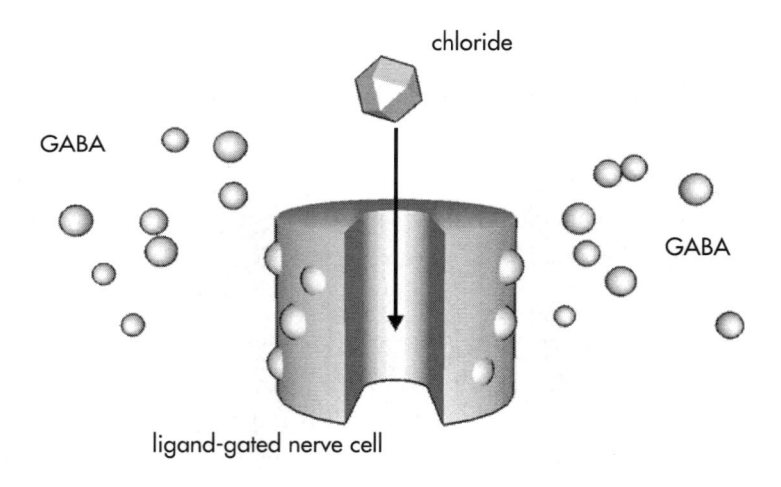

FIG. 1 : *GABA binding to a nerve cell allows an ion of chloride to enter it*

Throughout the body are neurons with *neural receptor sites* whose action modulates the response to stimulation. The receptor consists, among other things, of specialized cells whose function is to attract an inhibitory neurotransmitter the body manufactures called Gamma-Amino Butyric Acid, or GABA. What GABA does is to bind to parts of the neural receptors, in a process that causes the neural receptor to open, allowing an ion of chloride to enter (and an ion of potassium to exit.) When that happens, the electrical potential of the membrane is increased, which then counteracts any electrical stimulation of the neural receptor. It is this action that calms the nerves. Benzodiazepine binds to a different place on the neural receptor than where GABA binds, but its presence there strengthens the bond that the GABA makes, which increases the power of GABA to inhibit stimulation.

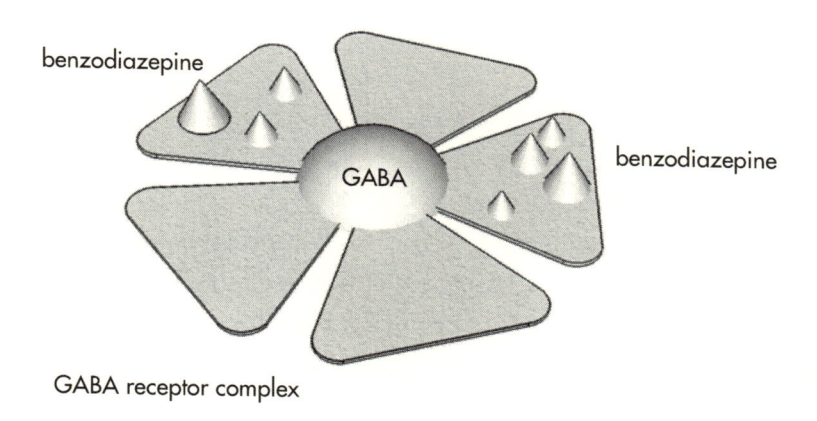

FIG. 2 : *Benzodiazepine binding to the alpha subunits of a GABA receptor*

What Professor C. Heather Ashton found in her extensive research was that, in some people, after exposure to benzodiazepines, the ability of the neural receptors to attract GABA is reduced. So, even after there is no longer any benzodiazepine in the body to influence the receptors directly, they still aren't able to bind enough GABA to themselves to inhibit electrical excitation sufficiently. This phenomenon is called 'down-regulation' of the GABA-α receptor sites.

Dr. Lance P. Longo and Dr. Brian Johnson of the American Academy of Family Physicians, wrote in *Benzodiazepines—Side Effects, Abuse Risk and Alternatives*, that:

> *With long-term high-dose use of benzodiazepine... there is an apparent decrease in the efficacy of GABA-α receptors, presumably a mechanism of tolerance. When high-dose benzodiazepines... are abruptly discontinued, this 'down-regulated' state of inhibitory transmission is unmasked, leading to characteristic withdrawal symptoms such as anxiety, insomnia, autonomic hyperactivity and, possibly, seizures.*

What Professor Ashton determined was that even when benzodiazepine isn't 'abruptly terminated' as described above, it still has the potential of down-regulating the action of the neural receptors to inhibit excitatory states. While this is a simple, uncomplicated difficulty, the function of GABA throughout the body is so widespread, so fundamental to the operation of a wide variety of bodily systems, that to have it impaired produces an opportunity for a vast array of possible problems to appear. What those problems are will be examined more fully in later chapters, but as an example, more neural receptors which utilize GABA are found in the gut than in the brain; so, one potential effect of benzodiazepine reduction or discontinuation is that serious difficulties in digestion or elimination of food may result. This would seem to have nothing to do with having used tranquilizers but to the body, relying as it does upon GABA for so many of its functions, it is yet one of the phenomena that may result from both tolerance to or discontinuation from benzodiazepine.

Other intoxicating agents have a similar effect on the body's ligand-gated neurons, *i.e.,* facilitating the effects of GABA, and some, such as barbiturates, not only influence GABA but supplant it by themselves causing the neuron to allow entry of chloride. They do not, however, bind as tightly, and more importantly, their presence doesn't affect the ability to attract GABA. **It is the down-regulation of the neural receptors that differentiates benzodiazepine dependency from that of** *all other substances.* That is the explanation of why getting off these drugs can be so horrendously difficult for some people, and why withdrawal symptoms for some can last days, weeks, months, *even years* after the last dose of benzodiazepine is taken. It is ignorance of this aspect of benzodiazepine discontinuation that leads medical professionals—even addiction specialists who should know better—to misunderstand the plight of benzodiazepine users. *All of their exceptional difficulties and often bizarre discontinuation phenomena are a result of the single problem of down-regulation of the neural receptors after exposure to benzodiazepine.*

Those reading this book who are in the throes of benzodiazepine dependency would do well to absorb that idea: the sole problem you are having is that benzodiazepine has interfered with one of your body's most elemental functions, that of attracting GABA to its neural receptors. The results of this condition may well present as a staggering array of withdrawal phenomena, everything from insomnia and anxiety, which would seem understandable, to dental distress, difficulty breathing, sinus problems, twitching muscles ... the list is seemingly interminable and contains phenomena that would appear to have nothing to do with the nervous system. Subjectively, these phenomena feel like *illness.* People quite naturally believe they are *sick* and require *healing.* But 'illness' normally implies either disease or tissue damage. What is happening to someone whose many trillion nerve cells are down-regulated because of benzodiazepine is neither disease nor tissue damage, but more like a mechanical malfunction. Therefore, recovery is more akin to 'repair' than 'healing.' Being aware of the actual nature of what is wrong helps demystify the rather bewildering process of getting on with recovery.

After the over-sedated period passed, I really began to feel the effect of lessening amounts of benzodiazepine in my system—which is to say, I was feeling the effects of nerves that did not have sufficient GABA to mediate stimulation properly. The first effect to present itself was insomnia, and it was awful. I could sleep no more than about three hours per night and went through each day with that bedraggled feeling that comes from not having had a good night's sleep. The torturous part was that all day I was so tired that it felt like the moment I put my head on the pillow I would fall immediately asleep. But when my head did hit the pillow, the effect was that I became suddenly instantly alert. Because of the heightened activity of my nervous system, when I would lie in bed and pull the covers up to my neck to keep warm the sound of the satin blanket binding rubbing against my neck seemed so loud I couldn't ignore it. I'd pull it down away from my ears to diminish the racket it made, but then my neck or shoulders would get cold. It was maddening. Every sound in the neighborhood would penetrate my cottage, even though the doors and windows were closed. I would feel my hands rustling against the covers and become aware of the sensation. Minor itches would demand to be scratched. I simply found it nearly impossible to fall asleep; and yet, if in frustration I finally got myself up and out of bed, the moment I was vertical I would feel exhausted, ready for bed and sleep.

To counteract the sense of torture I would feel when I lay down to sleep and found that it wasn't forthcoming, I began to tell myself, "Okay, I'm not falling asleep, but at least I'm resting. I'm resting my body, resting my eyes." That would help, but unfortunately, the mind, compromised as it was, never rested.

The problem of getting to sleep persisted for nearly three years. At one point, it even got worse, something I wouldn't have imagined possible. At that time, I had to devise a method for getting to sleep. At two or three in the morning, I would open a book and start to read. The book had to be utterly barren of anything that would interest me, so I would get old books from our local dump on subjects that I didn't care about, such as accounting methods in common use in overseas mining operations in the 1950's. I had learned the hard way that I couldn't use books that were interesting or else I would get too stimulated and stay awake all night, reading, even though my mind could not really comprehend the sentences. I would read, kneeling on the wood floor of my cottage. Hour after hour would go by while I read the dull pages, getting chillier by the minute, my knees and joints growing stiff and aching from kneeling on the cold floor. When I was finally perfectly uncomfortable, I would undress, and get into my bed beneath the covers. The sudden warmth of the blankets and softness of the bed would be so comforting in contrast to kneeling on the floor, that that would soothe me just enough so that I could fall asleep. The differential between the two states of comfort would trigger the sleep response—but if it didn't, I'd have to drag myself up out of bed and get back onto the floor to repeat the reading phase of the process.

What sleep I got was hardly refreshing. Instead of dreams, what I experienced while I slept would more accurately be described as 'disturbing images,' fugues of disjointed, imperfectly formed visual representations of grotesque people and bad situations, jumbled all together, underneath a soundtrack of multiple voices and noises, like being in a room with half a dozen radios, each playing a different talk-format station. When I would startle awake after three or four hours of this, I had no sense of having had a break from my own consciousness as sleep often provides; throughout the night, I was never unaware of myself.

One of the effects of Xanax—and all benzodiazepines, to differing degrees—is to disrupt Stage IV sleep, also called 'Delta Wave Sleep' or 'Slow Wave Sleep.' Disturbances in this vital portion of the sleep cycle are associated with all of the fatigue syndromes and are characterized by not feeling refreshed upon awakening.

In *Basic Pharmacology in Sleep-Related Disorders,* appearing in the Oct/Nov 2000 issue of *RT—The Journal for Respiratory Care Practitioners,* Thomas M. Kilkenny, DO, and Steve Grenard, RRT, wrote:

Benzodiazepines can affect the staging of sleep by increasing stage-II non-REM sleep and suppressing K complexes and stage-III or stage-IV sleep. The latency of REM sleep usually increases, but there is often a more frequent cycling of REM periods. The electroencephalograms of patients taking alprazolam will often show more numerous, denser, and lengthier sleep spindles. Movements during sleep are also suppressed. Chronic use of the medication can markedly decrease the amount of REM sleep, and abrupt discontinuation of a benzodiazepine will cause REM rebound, which is manifested by a marked increase in total REM sleep.

In my case, the "increase in total REM sleep" resulted in the nightmarish fugues I experienced every time I slept. I had taken Xanax for a total of about nine years, so my Stage IV sleep had been compromised for almost a decade. As a result, I had a massive amount of material in my subconscious mind that had never been processed in dreams and thus integrated with my consciousness. Evidently, there was vastly too much of this mental stuff to be sorted out and converted into cogent symbology by regular dreams, so it all came out in an interminable flood of images and dream sensations.

The effect was quite horrible. I craved sleep, and yet sleep was not merely unrestful and unrefreshing, it was characterized by the nightly horror show of dream images. This persisted for about two years; then, one night, the flow of disturbing images organized itself into a sequence of events rather than a multiplex of them. A few nights later, it happened again, then again, and within a week I was once more having regular dreams, with plotlines that followed the strange parallel logic dreams employ. I knew the ordeal of 'disturbing images' was over when I had a dream that reflected the events of the day; my mind had had restored the ability to process events and sensations as they occurred in my waking life. But that wasn't to happen until much later.

I avoided sensory stimulation as much as possible, continuing the state of being a housebound recluse that had begun when I was in tolerance withdrawal from Xanax. Any stimulation at all was too much stimulation. I was in a nearly constant state of anxiety, which would show up first as a fluttery feeling in my stomach, and then proceed to become a sort of icy clench that would seem to grip my heart. Adrenaline surged through my system almost constantly. Since this was my 'normal' state, I eventually devised somewhat homemade Cognitive Behavior Therapy skills to ignore it. I would tell myself, "I'm feeling fear and anxiety, but there's nothing in this immediate environment that can harm me, so these feelings aren't authentic. They're not real. They're the result of what's going on in my nervous system, that's all. They're chemical, not actual." I got where I could function in spite of the constancy of anxiety and adrenaline.

In this state, I interacted with other people only when it couldn't be avoided. Every two weeks I went to see my doctor to get my next prescription of Valium. He would ask how I was doing and I would attempt to communicate what I was going through, but he didn't seem to hear me. He didn't really get it. No one did. No one would, who hadn't experienced it. I looked fairly normal. No one could tell how hellish every second of every day was. I would have fared better with others if I had worn a blood-soaked bandage on my head, giving everyone a visual cue that I was brain-damaged; because that's what this was, brain damage. Hopefully, it was temporary, not permanent, but it was a form of chemical brain damage.

The lethargy and fatigue were so great that I never had any nervous energy at all. I would never be found jiggling my foot, or doing any of the myriad things that normal people do to express their bodily energy. When I sat in a chair or on the couch, I would sit there as though poured, unmoving. When I slept, I would lie on my right side and wouldn't move until I awakened some hours later, in the same position. There was no tossing or turning, or movement of any kind while I slept. I observed that when I walked, my arms hung straight by my sides. The muscles in my body settled into a rigid form, without flow or rhythm, just the mass that was my flesh.

About a month into my taper, a client came out to see me. He had tried to call me but since I had turned my phone ringer off, he could never get through. He had driven out to my house because his trouble was serious: his computer system had crashed and it appeared he had lost all of his business records. He very much needed my help.

I followed him back to his office in my car. Driving it was a challenge, to say the least. The car was a special two-seater Mustang with a powerful engine. If I pressed too heavily on the accelerator pedal, the engine would roar and the rear tires would squeal on the pavement. I have always loved fast cars but to drive such a vehicle requires a fully functioning nervous system. With my anxiety and in my frazzled state, the dynamic responses of my car would cause the already constant flow of adrenaline to surge even more than usual.

When I got to my client's office, he showed me to the computer and, fortunately, left me alone to work. I looked at his computer screen and saw immediately that his most important files were missing. "Okay," I said to myself, "I know I know how to fix this. I'll just have to remember what to do." Moments later, my thoughts were swimming in my head. "Wait a sec, what am I doing here, again? Oh, right. I'm here to restore the deleted files." In a progression of disjointed thoughts, I pieced together the procedures I would need to employ. Occasionally, I would lose the thread altogether and forget what I was doing and even why I was someplace other than in my cottage.

I knew that during the process of recovering the files, one wrong step could cause them to be lost forever, and I was well aware of the value of this data to my client. To attempt such a thing with my mind in such bad shape was daunting, but I knew it had to be done and that I had a responsibility to the client to do it. I finally figured out that if I wrote down the steps I needed to follow and then ticked them off as I performed them, I wouldn't have to worry about the brief mental lapses.

When it was all over and the data was safe, I left—without the client ever knowing how nearly impossible the task had been for me. Other interactions with people, ones that didn't require functioning at such a high level, were carried out in a sort of robotic autopilot mode I would lapse into. I would act out a sort of parody of myself, speaking and responding in the way that I historically would have behaved in such a situation. My actual consciousness would feel like a distant observer of the proceedings. Later, I would think, "Wow, that was pretty bizarre." And people would have had no idea what my true state had been while I had been interacting with them.

The incident of my first employment since I had been devastated by Xanax points out the strange circumstance that throughout the ordeal of getting off of tranquilizers I could always engage in rational thought. There was no doubt that I was extremely mentally unbalanced at times as well as intellectually diminished. Although I had formerly been proficient at dealing with abstract ideas, such abilities were beyond me, as was anything involving sensation, daydreaming, speculation, curiosity, inquiry—all of the ancillary dimensions of thought. What I was capable of was simple, reasoning thought itself. While I was most definitely in a condition of torment, I was never deranged or delusional. I never imagined seeing monsters in the shadows or thought I was Napoleon. I did, however, have the rather paranoid notion that I was the victim of a vast conspiracy by pharmaceutical companies—but then, that thought persists today.

Later, when I was capable of examining this peculiar phenomenon, it struck me that at the level of myself as merely an entity, not yet differentiated as the particular personality that I happened to be, I was always powerfully aware that my survival was very much at risk. Rational thought, the ability to figure out that, "if I do *this* then *that* will happen," is the human equivalent of fangs and claws. The part of myself most concerned with my survival in the face of the devastation that had happened to me held tenaciously onto my ability to think so that it could solve the problems that stood between me and the continuation of my existence.

Chapter Eight
Drugs and the FDA

In 1960, the Swiss-based pharmaceutical company, Roche, brought a new class of tranquilizer to market with their product, Librium. This was to be the first of the many benzodiazepines approved for medical use. Librium had been developed by a senior research chemist, Dr. Leo Sternbach, who worked for Roche's American branch, Hoffmann-LaRoche, Inc., based in Nutley, New Jersey. The company had relocated Dr. Sternbach—and other Jewish employees— from Switzerland to the United States during the early days of World War II, and after setting up the first laboratories in the American facility, he continued his work on the synthesis of vitamins. In 1954, Sternbach was charged with finding a drug Roche could use to compete with Miltown, a tranquilizer from a rival pharmaceutical firm, Wallace Pharmaceuticals. Among the substances he examined for potential use was benzheptoxdiazine, a chemical compound he had first worked on twenty years previously as a possible wood dye. He treated this substance with methylamine, which yielded a white crystalline powder, which he labeled 'RO 5-0690' and set aside for future study.

According to Roche corporate lore, a research assistant, cleaning Dr. Sternbach's laboratory in 1957, came upon RO 5-0690 and asked whether it should be thrown out. Sternbach pondered, then had the assistant send the compound to the company's pharmacological division for testing, where it was determined to have marked anxiolytic, *i.e.,* tranquilizing, effects on laboratory mice. Other accounts suggest that Sternbach made a pretence of having almost thrown out the compound, as Roche had ordered him to 'stop fooling around' with tranquilizer research and pursue antibiotics. Regardless, the new material, benzodiazepine, had shown that it was effective as a tranquilizing agent and Roche went to work to develop products based upon it. The first was Librium. It was found to have fewer side effects than Miltown, and became a successful competitor. Three years later, Sternbach and his associates had developed Valium, which was more potent—and less bitter—than Librium. The Age of American Tranquilizers had begun.

Valium became the first 'blockbuster' drug. During its peak year of sales, 1978, Americans consumed 2.3 billion Valium tablets. It was the single most prescribed drug in the country from 1969 until 1982, and was, according to a press release from Roche, "the largest-selling pharmaceutical in the world." Its sales helped make Roche a giant in the pharmaceutical industry. Having relinquished his rights to his employer, Dr. Leo Sternbach was paid a royalty of $1.00 on the patent in his name. He didn't like taking Valium himself, claiming it made him feel "depressed."

In spite of such widespread success, Valium began to come increasingly under fire toward the end of the 1970's. Elvis Presley's pill-popping and subsequent death was big news, and Valium was noted as having been one of the drugs in his system. Celebrity use of Valium was presented on a CBS *60 Minutes* exposé. The vast number of middle-class Americans on Valium began to be viewed with concern. In 1979, a United States Senate investigation was convened under the sobriquet, *Use and Misuse of Benzodiazepines,* where Senator Edward Kennedy of Massachusetts announced at its opening that *"[E]xcluding alcohol, diazepam is the number one drug problem in the U.S. today."* In Canada, a report on benzodiazepine use by Dr. Ruth Cooperstock led to parliamentary debates, which resulted in the reclassification of benzodiazepines as controlled substances. The United Kingdom was even more progressive in instituting reforms aimed at reducing the number of its population who were dependent upon tranquilizers.

But then something curious happened. Just as Valium was earning a reputation as a dangerous and overprescribed drug and its use began to diminish drastically, other variations of the benzodiazepine compound were unceremoniously introduced into the medical marketplace. These other formulations were distinguished by their seemingly unique attributes, most of which were a function of how quickly or slowly their effects were felt. And thus, some, such as Dalmane and Restoril, were touted as sleeping pills and others were offered as anti-seizure treatments or muscle relaxants, all this in addition to new preparations targeting anxiety, much as Valium had done. What is peculiar is that these drugs were basically the same thing as Valium; so if Valium were being excoriated, why wasn't that happening to these new drugs? It is akin to someone recognizing that beer can lead to health problems, but somehow believing that vodka, whiskey, gin and rum are okay. There is an intellectual disconnect in the acceptance that met the later benzodiazepines and it is still in force today. Pharmacies fill prescriptions for Xanax, Klonopin, Ativan and all the rest without raising an eyebrow, but grow suspicious if a customer hands them a prescription for Valium.

While differences in their therapeutic action seemed to make the second-generation benzodiazepines suitable for treating a far wider range of human ailments than just Librium and Valium alone, certain of those differences were so marked that they led to concern. One such drug was Halcion. Approved by the FDA in 1982, nineteen years to the day after it had approved Valium, Halcion became pharmaceutical company Upjohn's second best-selling drug, following only Xanax in sales. For a time, it was the most popular sleeping pill in the world.

Halcion had been introduced in Holland in 1977, but after two years on the market, its license was suspended for six months to examine claims that the drug was causing psychiatric disturbances in some of its users. As the result of an investigation, Dutch authorities permanently barred Halcion at doses higher than one-quarter milligram, twenty-five percent of the dosage originally approved for use in the country. Upjohn's response was to withdraw Halcion from the market, presumably to avoid having the efficacy of higher dosages questioned at a time when the FDA was in the process of reviewing the drug for release in the United States.

And yet, those questions arose. According to an article in the August 19, 1991 issue of Newsweek magazine, Dr. Theresa Woo had worked for the U.S. Food and Drug Administration on Halcion's application for use in the United States. Dr. Woo had contended that, based upon the evidence of the unfavorable patient responses in Holland, and upon Upjohn's studies as well, the drug should not be approved for use. The FDA approved it anyway. Halcion went to market in 1983. Woo had sought to temper the approval by requiring that Halcion's dose be no higher than one quarter milligram, as Holland had done, but had to reverse her position because studies had shown Halcion only to be effective at higher doses.

Many years later, long after Halcion was given license by the U.S. Food and Drug Administration to be prescribed to American patients, it came out that Upjohn had, in the words of FDA investigators, *"engaged in an ongoing pattern of misconduct"* during the application process. One of the studies crucial for the FDA's evaluation of the safety and efficacy of Halcion was 'PROTOCOL 321,' in which healthy male prisoners were given the drug and their reactions to it were observed. Thirty percent of the prisoners' unfavorable side effects were left out of the summary of PROTOCOL 321 when it was submitted to the U.S. Food and Drug Administration. Upjohn claimed that the omission was a result of 'transcription error' but that seems unlikely, as there were further omissions made as well. The side effects of two of the prisoners who participated in the study were grossly misrepresented, a fact that only came to light when the prisoners themselves were interviewed in the October 14, 1991 broadcast of the BBC program, *Panorama*. Interestingly, a representative of Upjohn, Dr. Robert Shaw, used a different study, 'PROTOCOL 6415,' to demonstrate the drug was safe. In fact, the data in PROTOCOL 6415 was found by the FDA to have been falsified. The doctor involved in the study, Dr. Samuel Fuerst, had submitted the names of patients who had reportedly used Halcion with no ill effects. It was subsequently discovered that these patients had never taken the drug; Dr. Fuerst had falsely concocted their data. The U.S. Food and Drug Administration disqualified Dr. Samuel Fuerst—along with three other investigators—in clinical trials for Halcion.

An important incentive for the suppression of data from PROTOCOL 321 was that it would have supported with scientific evidence the contention that Halcion not be prescribed for periods longer than fourteen days. While such a short term is considered medically prudent for benzodiazepine drugs, it would most certainly have cut into Halcion's profitability to Upjohn. An internal memo, which came to light during a civil trail over Halcion, warned that a two-week limit for use of the drug *"could reduce projected sales by 50% over a 10 year period."* The FDA report, which identified the distortions in PROTOCOL 321, concluded that the purpose of doing so appeared to have been to influence the FDA to set aside a proposed two-week limit *"even though available evidence indicated that long term use was both dangerous and medically untenable.... The firm chose to disregard the potential harm of inappropriate use, in order to gain additional sales (profits)."* The report itself was suppressed for many years, and only became available through requests made under the Freedom of Information Act.

Upjohn marketers need not have been overly concerned about use of the product being limited to short periods, however. *The Evaluation of Medications for Insomnia in Canada* reported that the average use of hypnotic agents was 1.7 years. As such drugs are only available from doctors, it is evident that doctors are quite willing to prescribe far outside the safety recommendations of the governmental agencies that approve their use.

The FDA report added that Upjohn conducted *"a continuous, ongoing campaign to discredit or neutralize any individual or publication reporting adverse information about Halcion."* A news item in the May 21, 1994 edition of the *British Medical Journal* reported: *"This included attempting to discredit and counter statements made by [Halcion critic] Professor Oswald; preventing publication of an article in the New England*

Journal of Medicine; supplying incomplete information to a conference; writing an *'inaccurate and misleading'* letter to the Lancet; and supplying *'incomplete and inaccurate'* information to drug agencies in France and Japan."

Such improprieties may not have convinced the FDA to keep Halcion out of the hands of the public, but they were enough to inspire the United Kingdom to do so. The British High Court's Medicine Control Agency revoked Upjohn's license to market Halcion in 1993. Upjohn appealed the decision to the European Court of Justice in Luxembourg in 1999, but its appeal was denied and the sale of Halcion is still banned in the United Kingdom.

The U.S. Food and Drug Administration's Center for Drug Evaluation and Research collects 'adverse reaction reports' from hospitals, physicians, health care providers and patients under a protocol known as the Spontaneous Reporting System. They are sent either directly to the FDA or given to the drug's manufacturer, who is required by regulation to send it on to the FDA. Adverse reaction reports are just that: reports of any unexpected or untoward response to a medication. Since the clinical trials which the FDA uses to determine the efficacy of new drugs are of necessity both short-term events and limited in scope, the FDA uses the Spontaneous Reporting System in order to assess the safety and efficacy of drugs once they are in the medical marketplace, by monitoring use over longer periods and by vastly larger numbers of patients.

After Halcion's release in the United States, adverse reports began to mount. Over the next few years, the U.S. Food and Drug Administration observed the adverse reaction reports filed through the Spontaneous Reporting System. The responses showed disturbing psychiatric side effects similar to the ones reported in Holland, including, according to the Newsweek magazine article, 'personality changes,' 'inappropriate emotional expression,' and 'unaccustomed aggression.' Such reports prompted a more extensive study by the FDA. The adverse reaction reports for Halcion were compared to those of Dalmane and Restoril, also benzodiazepines. A report submitted to the Division of Neuropharmacological Drug Products showed that there were 8 to 30 times more adverse reactions reported for Halcion than both of the other drugs, even though Dalmane and Restoril prescriptions outnumbered those of Halcion at that time.

The FDA conducted another study in 1989. It looked at adverse reaction reports of amnesia, anxiety, confusion, hostility, psychosis, and seizures related to Halcion use and compared them to Restoril. Once again, the quantity of such reports was greater for Halcion than Restoril—from 8 to 45 times greater. These data were provided to the Psychopharmacological Drugs Advisory Committee, but the committee concluded only that Halcion's label should be amended to include a warning that there may be a greater risk of the occurrence of amnesia than presented by similar benzodiazepines. Representatives from Upjohn had convinced the committee that adverse reaction reports were 'anecdotal evidence' and, as such, lacked scientific value. Since there were no clinical data to show the conditions that had been prominent in the material from the Spontaneous Reporting System, the Psychopharmacological Drugs Advisory Committee could not make recommendations beyond a *proviso* concerning amnesia. If anything, however, the discrepancy between the adverse reports and the clinical evidence should have suggested that the clinical studies were obviously poorly designed, and more thorough studies called for. One study that the committee relied upon, for example, cited only the *first* side effect reported by its subjects. Therefore, if a patient woke up from sleep feeling groggy from the dose of Halcion the previous night, and then became paranoid later in the day, only the grogginess would be reported.

In 1997, because of the controversies around Halcion, the Institute of Medicine (IOM) produced, *Halcion, An Independent Assessment of Safety and Efficacy Data*. A committee had been empanelled to examine more than twenty years of data about the drug, from the studies used in evaluating Halcion for FDA approval through adverse reports from the Spontaneous Reporting System and post-marketing drug data. The report concluded that Halcion was safe and efficacious within the current labeling guidelines. It further concluded that,

The data from premarketing clinical trials, postmarketing studies, and the published literature do not support clearly the existence of a unique profile or syndrome of adverse events associated with Halcion relative to those associated with other drugs of its type.

The IOM report went on to say,

It seemed that at least some of the adverse events that were being reported through the Spontaneous Reporting System of FDA were similar to those that had been reported in some of the early clinical trials with higher doses and longer durations of use of Halcion. This, combined with survey data that indicate that many people use hypnotic agents for very long periods of time ... led the committee to consider the possibility that the adverse events that were being reported for Halcion might be due, at least in part, to the use of Halcion for longer periods of time and at higher doses than those currently recommended in the labeling.

While, superficially, the Institute of Medicine appears to have discovered that the cause of the phenomena revealed in the adverse event reports was overdosing, there is something wrong in the logic of their conclusion: if there is no "unique profile or syndrome of adverse events associated with Halcion relative to those associated with other drugs of its type," then it would not have generated such a disproportionate quantity of adverse events reported compared with Dalmane and Restoril. In other words, both those other benzodiazepine-based sleeping pills would have been equally misprescribed for longer periods and at higher doses than their labels recommended, which would have resulted in an equal number—and type—of adverse reaction reports being received through the Spontaneous Reporting System. That Halcion was responsible for many times the number of such reports than Dalmane and Restoril combined shows evidence of something unique about it that the distinguished scholars and doctors of the Institute of Medicine somehow failed to recognize.

A quick search on the Internet about Halcion yields this list:

• *More common side effects may include:*

Coordination problems, dizziness, drowsiness, headache, lightheadedness, nausea/vomiting, nervousness

•*Less common or rare side effects may include:*

Aggressiveness, agitation, behavior problems, burning tongue, changes in sexual drive, chest pain, confusion, congestion, constipation, cramps/pain, delusions, depression, diarrhea, disorientation, dreaming abnormalities, drowsiness, dry mouth, exaggerated sense of well-being, excitement, fainting, falling, fatigue, hallucinations, impaired urination, inappropriate behavior, incontinence, inflammation of the tongue and mouth, irritability, itching, loss of appetite, loss of sense of reality, memory impairment, memory loss (e.g. traveler's amnesia), menstrual irregularities, morning 'hangover' effects, muscle spasms in the shoulders or neck, nightmares, rapid heart rate, restlessness, ringing in the ears, skin inflammation, sleep disturbances including insomnia, sleepwalking, slurred or difficult speech, stiff awkward movements, taste changes, tingling or pins and needles, tiredness, visual disturbances, weakness, yellowing of the skin and whites of the eyes

The obvious question that arises upon considering the actions of the FDA is, why would the agency be so supportive of a drug that other countries—the United Kingdom, Brazil, Argentina, Norway and Denmark—have banned due to safety concerns? If it is the role of the U.S. Food and Drug Administration to protect American citizens from such harmful drugs, why would it put the interests of a pharmaceutical company over those of the people?

The short answer is, naturally, *money.* The pharmaceutical industry is vastly wealthy. In 2004, its global revenues were $550 billion, of which $235.4 billion, roughly 43% of worldwide revenues, was generated in the United States. Pharmaceuticals are a vastly profitable industry, as well. According to Neal Pattison and Luke Warren of the watchdog group, Public Citizen, the ten largest pharmaceutical companies in 2002 had a median profit margin of 17%. That is a staggering amount, considering the fact that the average profit margin for all other Fortune 500 industries is 3.1%.

An industry that rich can afford to spend significant amounts of money in courting power; in fact, by doing so, it can create the opportunities for making even more money. The magnitude of the pharmaceutical industry's investment is considerable. According to Public Citizen: "*The drug industry's political spending came to more than $230 million for the 1999–2000 election cycle. This record amount includes: approximately $170 million for lobbying; almost $15 million in direct campaign contributions; at least $35 million in campaign ads by the drug industry front group, Citizens for Better Medicare; and $10 million funneled to the U.S. Chamber of Commerce for pro-drug-industry campaign ads.*"

Such investments pay off. Legislators whose campaigns have been underwritten to so large a degree by pharmaceutical companies have a vested interest in passing laws favorable to those companies. As an example of how this works, in 2003, Senator Richard Burr presented a bill called *Project Bioshield Act of 2003* which ostensibly was intended to protect the American public by providing for a partnership between the government and the pharmaceutical industry for the purpose of developing drugs and vaccines rapidly in case of a bioterror attack.

While that certainly seems both reasonable and desirable, the bill—and its successor, the *Biodefense and Pandemic Vaccine and Drug Development Act of 2005,* known as *Bioshield II*—contain Draconian provisions. Under Bioshield, the government can require all U.S. citizens to receive a vaccination whether they want one or not. Further, the bill makes provision for an entity within the government called the 'Biomedical Advanced Research and Development Agency' (BARDA), which, among other things, would work with pharmaceutical companies to develop medical countermeasures for both biological weapons and natural disease outbreaks. Within the provisions of the bill itself, BARDA is to be granted complete exemption from disclosure under the Freedom of Information Act, and such disclosures are not even subject to judicial review by courts of law. Pharmaceutical companies can produce drugs and vaccines to be given to all U.S. citizens without provisions for the public to determine what such drugs and vaccines are. 'Informed consent' would be, therefore, completely obviated as the public would be given these substances without regard for consent, and without being informed as to what the substances are. Should any citizens be harmed by them, exemption from the Freedom of Information Act will keep the public from ever even learning about it. Prohibited by the terms of Bioshield from constitutionally guaranteed legal recourse, victims of the drugs and vaccines would not be allowed to sue pharmaceutical manufacturers, even if it could be shown that those companies violated drug safety laws or engaged in negligent or fraudulent activities in producing them. Drug companies would be legally exempt from liability.

Contained within this extraordinary legislature is an equally extraordinary provision. It allows pharmaceutical companies to extend their current patents on any drugs deemed to be classed as countermeasures to biological weapons, even drugs that would appear to have no direct applicability to specific bioterror threats. BARDA would possess the entire authority to make such a determination and its decisions would not be subject to judicial oversight.

The big pharmaceutical producers lose enormous profits when their blockbuster drugs reach the end of their patent protection, at which time other companies begin to offer generic versions of the same drug at often greatly reduced cost. The second largest-selling drug in the world in 2005 was Zocor, a statin drug manufactured by Merck, the source of one-fifth of the company's earnings. Zocor's patent expiration in 2006 would result in Merck's annual revenues from the drug dropping by $2 billion. Industry analysts warn that losses due to drugs coming off patent protection will amount to $80 billion by 2008.

The inclusion in Bioshield of a provision for pharmaceutical companies to extend their patent protection represents literally billions of dollars to the top corporations in the industry. Who would champion such an anti-consumer measure in the Bioshield Project Act? Its sponsor, Senator Richard Burr, received $297,934 as campaign contributions from pharmaceutical companies in 2004, as reported by The Center for Responsible Politics. Co-sponsor of the bill, Senator Bill Frist, received $260,373 from them for the 2000 Senate race in his home state of Tennessee. *Quid pro quo.*

Drug companies use less direct methods to influence those in government outside of the Capitol building, such as the policy-makers at the U.S. Food and Drug Administration. Financial incentives abound in and around a wealthy and highly profitable industry and employment is well compensated. The so-called 'revolving door' between the FDA and pharmaceutical companies serves the purpose of making certain that the interests of the major drug producers are well represented in the agency. After working at the FDA long enough to have established important connections, many people leave the administration and are hired by pharmaceutical companies at a considerable pay increase. Even more brazenly, some executives go right from drug companies into influential positions at the FDA, only to return to the drug companies later at higher salaries.

This has created a 'pro-industry' atmosphere at the Food and Drug Administration. Dr. Richard Burroughs, in an interview in the July/August 1991 issue of *Eating Well,* described the shift in atmosphere during his ten years at the Center for Veterinary Services division of the FDA: *"There seemed to be a trend in the place toward approval at any price. It went from a university-like setting where there was independent scientific review to an atmosphere of 'approve, approve, approve.'"* Like many other scientists, Dr. Burroughs was squeezed out of the agency for trying to block approval of drugs which pose a threat to safety.

Another such scientist, Dr. Michael Elashoff, PhD, worked as a biostatistician at the Food and Drug Administration from 1995 to 2000. A small biotech firm in Australia, Biota Holdings, had developed a drug called *zanamivir* which showed promise as an antiviral medication. Lacking the 'pipeline' to introduce their discovery, Biota Holdings contracted with Glaxo Wellcome Pharmaceuticals to use their considerable resources to market the drug throughout the world. Glaxo Wellcome (now known as GlaxoSmithKline) presented the drug to the FDA for approval. Given the trade name *Relenza,* zanamivir looked like an effective drug for treating two common strains of influenza. In practice, however, the drug could, at best, reduce the duration of a bout of the flu by only 1.5 days, not conferring much of a benefit to justify the cost of using it. But there was more.

It became apparent to FDA researchers studying the drug that it also posed serious safety concerns. Patients who have asthma or other respiratory ailments were found to be at risk of experiencing bronchial spasms, a potentially life-threatening condition. Since the FDA's approval process is meant to balance the potential benefit of a drug against the potential risk, a drug that gambled death in order to provide merely about a day less of flu symptoms seemed hardly worth the risk.

In February of 1999, reviewers on the Antiviral Drugs Advisory Committee at the Food and Drug Administration voted 13 to 4 to reject its approval. During the review, one committee member, Dr. John D. Hamilton of Duke University, said, *"There isn't sufficient efficacy to warrant me recommending this drug for my family or myself.... I just don't think it has sufficient effectiveness."* And yet, the FDA subsequently approved the drug anyway for use by Americans. Shortly after, the FDA had to issue a rare 'update' to warn that Relenza *"has not been shown to be effective—and may carry risk—in patients with severe asthma or a lung condition called chronic obstructive pulmonary disease."* Use of the drug has been implicated in twenty-two deaths due to breathing complications.

Dr. Elashoff was one of the FDA personnel who presented data at the meeting of the Antiviral Drugs Advisory Committee, where he pointed out that the statistics revealed that the efficacy of Relenza had not been demonstrated. In an interview for PBS's November, 2003 'Frontline' program, *Dangerous Prescription,* he reported that:

> *The next day after the advisory committee, several people in FDA management told me that they blamed me for the drug getting turned down in the advisory committee; that I wouldn't be allowed to present at the advisory committee meetings in the future for any other drugs.*

He was 'marginalized' and, eventually, quit in frustration, saying:

> *The FDA has a big problem. It's pervasive. It's throughout the entire FDA review culture. I didn't see it getting any better, which is ultimately why I left. . . .*
>
> *I think it was pretty well understood that if you were advocating turning a drug down—particularly if it was from a large pharmaceutical company—that that wouldn't be good for your career, as far as promotions. It wouldn't be good for your career, scientifically, as far as being able to review other drugs in the future that had potential problems.*
>
> *There is no room for bad news, particularly when large pharmaceutical companies are involved. I think with smaller biotech companies, particularly where it's their first drug application, I saw a little more intellectual honesty about the pros and cons of drugs. But for large pharmaceutical companies, it was pretty clear. I mean, it's called the drug approval process. It's not called the drug review process. So that really sets the mindset on what the job is.*
>
> *The people who stay for the long term are those who aren't unduly*

upset about the fact that drugs are getting approved that shouldn't be, or that reviews are being influenced either by drug companies or FDA management.

The whole promotion environment is such that people who raise concerns about drugs don't get promoted. So you have a whole set of people at the top who probably didn't have any morale problems, because they didn't see what they were doing as anything different from what they were supposed to be doing.

The ones who had ethical concerns—there's no reason to stay around in an environment such as that for year after year, when it's really so hard to make a difference. So those people would leave, and the ones who stayed might think this is how the drug approval process is supposed to go....

I think proof that people are being exposed to unsafe or ineffective drugs comes when drugs are pulled off the market for problems, where in most cases, warning signs were seen at the earliest stages of the review. It would just take either more people to die after taking medication, or just such a public recognition of the fact that a particular drug just wasn't effective or had many safety problems. When those drugs are pulled off the market, that's only the tip of the iceberg, as far as what other drugs are causing similar problems that there's just not an awareness of yet.

The influence of the pharmaceutical companies on the FDA is extensive. Dennis Cauchon, in an article in the September 25, 2000 edition of USA TODAY newspaper, wrote:
According to a USA Today study, more than half of the experts

hired to advise the government on the safety and effectiveness of medicine have financial relationships with the pharmaceutical companies that will be helped or hurt by their decisions. These experts are hired to advise the Food and Drug Administration on which medicines should be approved for sale, what the warning labels should say and how studies of drugs should be designed. The experts are supposed to be independent, but USA TODAY found that 54% of the time, they have a direct financial interest in the drug or topic they are asked to evaluate. These conflicts include helping a pharmaceutical company develop a medicine, then serving on an FDA advisory committee that judges the drug.

Federal law generally prohibits the FDA from using experts with financial conflicts of interest, but according to the article, the FDA has waived the restriction more than 800 times since 1998. These pharmaceutical experts, about 300 on 18 advisory committees, make decisions that affect the health of millions of Americans and billions of dollars in drugs sales. With few exceptions, the FDA follows the committees' advice.

The FDA reveals when financial conflicts exist, but it has kept details secret since 1992, so it is not possible to determine the amount of money or the drug company involved.

The shift in the FDA which resulted in its rubber-stamping pharmaceutical companies' products was initiated not for venal but, rather, for humanitarian reasons. During the first ten years of the burgeoning AIDS epidemic in the United States it was felt that the process by which medicines were painstakingly reviewed before approval by the FDA was too lengthy. People with AIDS were dying for the lack of effective treatments, some of which were believed to be imminent but stuck in a bottleneck of bureaucracatic red tape at the Food and Drug Administration. In response, President Bill Clinton instituted in 1992 a change in policy at the FDA to fast track approval for AIDS and cancer drugs. Pharmaceutical companies were quick to take mercenary advantage of the new policy, however, and welcomed the opportunity to get, not just AIDS and cancer drugs, but *all* of their newly-developed products approved and out into the marketplace more quickly.

The examples of the machinations surrounding the marketing of Halcion and the review process for Relenza reveal the cross purposes at which the medical establishment sometimes works. Pharmaceutical companies are not humanitarian organizations. They are corporate entities, whose single purpose is to return profits to their shareholders. Ultimately, all of their actions are bent to that task, and they have shown themselves to be remarkably successful at it. Half a trillion dollars in worldwide sales volume annually is a considerable amount, especially considering the sobering fact that the drugs they produce— with the exception of antibiotics—don't actually cure anything. Pharmaceutical drugs are almost always palliative, they 'treat' rather than 'cure' diseases, the result of which is that the consumer continues to buy them because the condition for which they were prescribed is still present.

It is this constellation of factors—doctors who prescribe drugs far beyond the limits recommended for safe usage, pharmaceutical companies that downplay the dangers their products present, and governmental regulatory agencies that sometimes tend to serve the needs of industry over those of the citizenry—that conspires to create the mindset of 'modern medicine.' Having been brought up in a simpler time in which medical people were believed to be incapable of doing us harm, most of us place ourselves in the hands of health-care providers with unthinking trust. Unfortunately, that is no longer an appropriate attitude to take.

Consider this. We are all suitably frightened of heart disease and cancer. Statistics appear to show they are the number one and number two causes of death in the United States. Media reports keep us continuously aware of these threats, but they never mention the third leading cause of death: *medical treatment itself.*

In the July 2000 issue of the *Journal of the American Medical Association,* Dr. Barbara Starfield, of the prestigious Johns Hopkins School of Hygiene and Public Health, wrote an article in which she analyzed data on mortality in the United States from studies of hospitalized patients. She found that 12,000 deaths were caused by unnecessary surgery, 27,000 were caused by medication and other errors in hospitals, 80,000 were caused by infections in hospitals, and a staggering 106,000 deaths were caused by 'non-error, negative effects of drugs.' This yields a total of 225,000 annual deaths 'from a physician's activity, manner, or therapy,' trailing only heart disease and cancer as causes of mortality.

The following year a report emerged called, *The American Medical System Is The Leading Cause Of Death And Injury In The United States,* by Gary Null, PhD, Carolyn Dean, MD, ND, Martin Feldman, MD, Debora Rasio, MD, and Dorothy Smith, PhD. The authors had reviewed thousands of medical papers and collated the statistics into one study. They wrote:

> *The number of unnecessary medical and surgical procedures performed annually is 7.5 million. The number of people exposed to unnecessary hospitalization annually is 8.9 million. The total number of iatrogenic [induced inadvertently by a physician or by medical treatment or diagnostic procedures] deaths is 783,936.*

> *The 2001 heart disease annual death rate is 699,697; the annual cancer death rate is 553,251. It is evident that the American medical system is the leading cause of death and injury in the United States.*

In June of 2000, the British Medical Journal reported that, during a strike in Israel that year by physicians in public hospitals which had imposed sanctions on the procedures they would perform, the death rate dropped by a considerable amount. In an interview, Meir Adler, who manages a funeral parlor in Jerusalem, stated, *"There definitely is a connection between the doctors' sanctions and fewer deaths. We saw the same thing in 1983 [when the Israel Medical Association applied sanctions for four and a half months]."* In Colombia, the *National Catholic Reporter* reported in 1976 that, during a 52 day strike by doctors, the mortality rate decreased by 35%.

It is not to be suggested that physicians are incompetent, nor are they greedy, money-grubbing parasites who get rich from knowingly dispensing poisons. Doctors believe in the drugs they use to treat us, and we believe in our doctors. It is important, however, to be grounded in reality rather than in unfounded beliefs, and the reality of the current health care situation is that it holds far more dangers for us than we could ever have imagined. Since the media, both print and television, are dependent upon lucrative pharmaceutical advertising contracts, it is unlikely that news about the venality of the pharmaceutical industry will ever get much coverage, or news that might shake the trust the populace has in our governmental agencies—or our health care providers. But the reality is that the entire medical system is quite fallible. And that is how it is that a drug which carries with its use the potential of creating dependency and devastating side effects came to appear on my doctor's prescription pad.

Evidently, we must learn to take responsibility for our own health, not delegate authority over our bodies to others. We need to view health care providers as facilitators who help us manage our health, rather than masters of it. We must fully empower our informed consent, and take the responsibility for doing our own research into the risks and benefits of what we choose to put into our bodies.

Chapter **Nine**
Withdrawal Symptoms

I had completed the first phase of the tapering process, reducing the amount of my daily intake of diazepam by 5 milligrams each week. Having started out at a dosage of 60 mg daily, I had cut that down to 40 mg in four weeks' time. To reduce from there, however, I could no longer decrease the dosage by 5 mg each week, as that would have represented too great a reduction. With that significantly lower amount of benzodiazepine present at my body's neural receptor sites, their ability to mediate nerve impulses would be too compromised, causing system-wide stress which might well traumatize me further.

A cut of 5 mg reduces a dosage of 60 mg by 8.33%. At each successively lower dosage, though, 5 mg constitutes an ever-larger percentage of the amount of the drug. At 40 mg., the same 5 mg cut then would be 12.5% of the dose. It appeared that 10% was about the maximum a cut should ever be, so when I continued past the 40 mg mark, I reduced my dosages by 1 mg per week rather than 5 mg. A 1 mg reduction is only 2.5% of 40 mg., so the impact on my nervous system of such a cut would be minimized.

Any yet, although these cuts were minimized, I began to feel their impact nonetheless. With the long half-life of Valium it took a few days for the lowered dosage to be perceived by the body, but when it happened, there would be an increase in discomfort. As I was already quite 'uncomfortable,' the cut just added to the distress I was feeling.

Benzodiazepines are widely-known for their disruption of memory. In fact, one of the most common uses of Xanax is for surgical procedures: the drug's anxiolytic properties help reduce patients' anxieties about undergoing surgery, while its *amnestic,* or 'amnesia-causing,' effect can diminish their memory of the unpleasant aspects of the procedures themselves, thus reducing the potential for post-traumatic stress. Benzos appear to interfere with *episodic memory,* while leaving *semantic memory* relatively intact. *Semantic memory* is the recollection of generalized information about the world, untied to the specific events that occurred as the knowledge was encoded. *Episodic memory* is the ability to recall one's personal experience of actual events. Benzodiazepines can signficantly disrupt the memory of such events, which would explain their use as 'date-rape' drugs. Victims who have been surreptitiously given such a drug, most typically Hoffman-LaRoche's *flunitrazepam (Rohypnol),* tend to develop anterograde amnesia about specific incidents which may have occurred while they were on it. Thus, they may have only vague, distorted recollections of sexual encounters, and their inability to recall specific details would render sexual assault under such circumstances of dissociative intoxication difficult to prosecute.

During the time when I was taking Xanax daily, my own memory deteriorated remarkably, although I never suspected that it was related to my use of a tranquilizer. When my son was fifteen years old, he happened to mention to me that he tended to have high blood pressure. "How can you have high blood pressure?" I protested. "You're a fifteen year-old kid! How do you even know what your blood pressure is?"

"They took my blood pressure when I had pneumonia, Dad," he replied.

I was somewhat shocked. "You had pneumonia?"

"Yeah, Dad, don't you remember? I stayed home from school for three weeks and you took care of me. You made me soup every day."

"When did this happen?" I asked.

"Two months ago."

I was appalled. How could I could have forgotten such a momentous thing? How could I have forgotten—after only two months—an 'event' I had participated in which had stretched out over an extended period of three weeks?

Fortunately, as the level of benzodiazepine in my system decreased during the tapering process, the drug's proclivity for disrupting episodic memory diminished accordingly, and my ability to recall specific events slowly returned. In fact, I began to have memories of events which occurred in my childhood and young adulthood, long before my period of Xanax use. It made me realize that, during that period, I had operated in a way that was disconnected with the normal sense of continuity that someone's life tends to have, rather like a chemically-induced version of 'living in the moment.' I could only view such disconnection as psychologically unhealthy, and was grateful that the slow withdrawal from the agent which was causing it was reestablishing my ability to recall the myriad events that had happened to me, giving me back a sense of the *context* in which my life evolves.

As the dosage reduction continued, I also began to develop 'w/d's', as they are abbreviated, or *withdrawal symptoms*. The general malaise I was experiencing was, itself, a withdrawal symptom as it was a result of not having enough benzodiazepine available in my system; or, more accurately, it was a result of not having enough GABA binding to my neural receptors as a result of those neurons having been exposed to benzodiazepine over a period of time. The overall effect of a down-regulated GABA system was manifested in pervasive fatigue, an inability to sleep, and an inability to have any positive feelings whatsoever.

Normally an optimistic, enthusiastic, somewhat cheerful person, I became instead absolutely wretched and miserable. My character certainly didn't change as a result of having taken a drug. Optimism and enthusiasm that arise from a person's consciousness, however, require certain neurochemical events to take place within the body in order to take expression. In my condition those events were not possible, evidently. The effect was that my moment-to-moment experience of myself was one of utter and abject misery, which varied only in the degree to which I felt miserable. Interestingly, I never forgot what it was like to feel good, even though actually feeling good and feeling happy became neurologically impossible for me.

When I was in the company of others I tried as much as possible not to impose my own condition on them. Instead of presenting myself as I actually felt myself to be, I developed the ability to project my 'historical' self, *i.e.,* myself as I *normally would have been*. It was rather like being an actor, but the character I was portraying was my own self. My memory of how I would normally behave, the things I would normally say, was complete enough for me to sustain rather lengthy improvisations, limited only by the fact that it took energy to interact with others. While I was focused on these interactions I would often forget how I was feeling—only to have it all come crashing back in on me once I was again alone.

Aside from these constant symptoms of withdrawal, there was a host of other w/d's that developed as my dosage of Valium crept ever lower. Persistent dizziness, which waxed and waned in intensity, required me to move slowly and conscientiously at all times so that I wouldn't lose my balance. The intense, nearly total anxiety I had experienced after making my experiment with Effexor had diminished. In its milder form I would feel the classic 'butterflies in the stomach' sensation, or it would escalate from that into a sensation like icy fingers gripping my heart in my chest. The anxiety phenomena appeared to operate independently of the adrenaline surges, which were almost constant.

I developed a strange shortness of breath that would arise whether or not I had tried to exert myself. Later this would be accompanied by occasional wheezing. Both seemed to be related to sinus problems, which also began as my dosage decreased. I had not only an excess of mucus but an excess of saliva as well. At times the saliva would have a sweet, metallic taste, which would persist for days, then vanish. When it got closer to my bedtime, an electric sensation would begin in my feet and run up my legs. It would increase in intensity until I was squirming around in my chair, trying in vain to relieve it by movement, by flexing the muscles. Oddly, if I waited until these symptoms were at their peak, at an almost unbearable level of discomfort, and then got into my bed, I would almost invariably fall asleep almost immediately, far more quickly than the nights when the Restless Leg Syndrome was absent. It seemed incredulous that I could possibly fall asleep with my legs writhing in my bed—yet it happened time and time again.

One of the most annoying of these w/d's were *fasciculations,* or twitches of small muscle groups. Random muscles in my legs and sometimes my forearms would spasm rhythmically, persistently. This phenomenon would occur either sporadically or almost constantly. At times, I would look down at my legs and the muscles were firing in so many places it would appear that there were snakes crawling around under the skin.

As maddening as it was to experience such things, it would have been far more maddening if I hadn't known they were sequellæ of my reduction of benzodiazepine. There seemed nothing to connect such bizarre physical ailments to drug usage, but fortunately for me, there was ample information about such things in the combined knowledge available at *benzo.org.uk*. Not only had Ray Nimmo assembled everything known about benzodiazepine—from news stories to published scientific papers—but there were the personal accounts of numerous individuals who had gone, or were going through, the process of withdrawal. In the bulletin board section of the website people would report on their experiences, so if someone shared an account of a physical problem they were having, there would be others to respond instantly if that same problem were something they, too, had contended with.

And this was a fortunate thing, as the number of reported withdrawal symptoms is as staggering as the nature of them is incredible. Hit with a multiplicity of such often quite bizarre phenomena, most people would respond, understandably, by going to their doctors. Doctors, however, seem largely ignorant about there being any extraordinary difficulties in discontinuing benzodiazepine drugs, since most of their patients who have undergone the process evidently did so without incident. There are few means for doctors to learn about the problem, either, because so few scientific studies exist to alert them to the dangers benzodiazepines pose. (There is no economic incentive to support such studies, as their outcome could only result in selling fewer drugs, while the medical marketplace demands the continued expansion of drug sales.) Most doctors got their fundamental knowledge about drugs during their years at medical school, but their main source for information about newer drugs is from drug company sales representatives. Lacking any medical training, sales reps present only such information as will accomplish their function—to sell their companies' wares—yet doctors accept their pronouncements as though they were objective and thorough assessments. Sales personnel are hardly likely to suggest to doctors that there are dangerous discontinuation phenomena associated with the particular benzodiazepines they sell because to do so would result in lowered sales, and therefore, lowered sales commissions.

As a result, when patients present their strange problems to their physicians, the response is usually to order tests to look for an underlying pathology causing the symptoms, and to issue more medications to treat them. That leads to further complications, naturally, as the new medications will have their own impact on the processes that occur while someone is affected by benzodiazepine. If it were understood, however, that the essential problem caused by the drug is that, in some individuals, it diminishes the body's ability to utilize GABA, even after discontinuation, many of the complaints would then make sense. Anxiety and insomnia are two more common of these complaints, but they are easily recognized as understandable problems that would result from having become habituated to a drug prescribed for anxiety and insomnia. Something like the fasciculations I experienced, however, would hardly seem to be a result of tranquilizer use.

And yet, there is a logical connection. Benzodiazepines are often prescribed as muscle relaxants. For a period of almost ten years, therefore, my body grew accustomed to maintaining muscle tone while exposed to a chemical compound that has a marked antispasmodic effect. In the absence of that chemical, the body could no longer manage muscle tone, and fasciculations, *i.e.,* muscle spasms, resulted. In others, this circumstance has led to muscle tensions, most usually a tightening up of the muscles in the shoulders and neck.

While some withdrawal symptoms can be understood logically, though, others seem to have no connection at all with the influence of a tranquilizer. The website for TRANX in Australia lists these symptoms as effects of benzodiazepine use or discontinuation:

Common withdrawal symptoms: abdominal pains and cramp, agoraphobia, anxiety, breathing difficulties, blurred vision, changes in perception (faces distorting and inanimate objects moving), depression, distended abdomen, dizziness, extreme lethargy, fears, feelings of unreality, flu-like symptoms, heavy limbs, heart palpitations, hypersensitivity to light, indigestion, insomnia, irritability, lack of concentration, lack of co-ordination, loss of balance, loss of memory, muscular aches and pains, nausea, nightmares, panic attacks, rapid mood changes, restlessness, severe headaches, shaking, seeing spots before the eyes, sore eyes, sweating, tightness in the chest, tightness in the head

Less common withdrawal symptoms: aching jaw, craving for sweet food, constipation, depersonalisation (a feeling of not knowing who you are), diarrhoea, difficulty swallowing, feeling of the ground moving, hallucinations (auditory and visual), hyperactivity, hypersensitivity to sound, incontinence, or frequency and urgency, increased saliva, loss or changes in appetite, loss of taste or metallic taste, menstrual problems (painful periods, irregular periods, cessation of periods), morbid thoughts, numbness in any part of the body, outbursts of rage or aggression, paranoia, painful scalp, persistent, unpleasant memories, pins and needles, rapid changes in body temperature, sexual problems (changes in libido), skin problems (dryness, itchiness, rashes, slow healing), sore mouth and tongue, speech difficulties, suicidal thoughts, tinnitus (buzzing or ringing in the ears), unusually sensitive (unable to watch the news on television or read the newspaper), vaginal discharge, vomiting, weight loss or gain

Rare withdrawal symptoms: blackouts—an episode where the person has no recall of their activity (this is rare with low dose use, but less rare when large amounts have been taken), bleeding from the nose, bleeding between menstrual cycles, burning along the spine, burning sensation around the mouth, discharge from the breasts, falling hair, hæmorrhoids, hypersensitivity to touch, rectal bleeding, sinus pain, seizures (fits) (almost unknown if people reduce gradually, more common for people using high doses who stop suddenly), sensitive or painful teeth, swollen breasts

An even more comprehensive list can be found at Ray Nimmo's website, *www.benzo.org.uk.* These symptoms have been collected from people who have reported having them both during and after discontinuation. From the TRANX list above, I, myself, have experienced:
agoraphobia, anxiety, breathing difficulties, blurred vision, depression, distended abdomen, dizziness, extreme lethargy, heart palpitations, hypersensitivity to light, insomnia, lack of concentration, lack of co-ordination, loss of balance, loss of memory, muscular pain, nausea, nightmares, craving for sweet food, diarrhoea, difficulty swallowing, hypersensitivity to sound, increased saliva, metallic taste, morbid thoughts, numbness in the body, sexual problems (changes in libido), speech difficulties, suicidal thoughts,

buzzing in the ears, unusually sensitive (unable to watch the news on television or read the newspaper), weight gain, hypersensitivity to touch, sinus pain

To doctors and addiction specialists, accustomed as they are to the withdrawal symptoms of people coming off of opiates, stimulants, alcohol and other drugs, the idea that these phenomena might be caused by benzodiazepines would seem highly unlikely and, perhaps, fantastic. To understand the connection, it is important to understand the relationship between benzodiazepine and how GABA functions in the body.

The body houses a Central Nervous System (CNS) contained within the brain and spine, and the Autonomic Nervous System (ANS), whose neural pathways extend throughout the body. Both systems are comprised primarily of *neurons*. These are specialized cells which conduct electrical impulses and it is estimated that there are 100 billion such cells in the human brain, with an equal amount in the spinal cord and the digestive tract. The electrical impulses convey information throughout the system, and this flow of information is controlled by *neurotransmitters* which affect how the neurons function. There are *excitatory neurotransmitters* that increase the electrical activity of the neurons and *inhibitory neurotransmitters* which reduce it, limiting the transmission of information. A critical balance between these two types is crucial: too much inhibition of neurotransmission would lead to a depressed, sedated condition, while too much excitation can cause people to feel anxious, unable to sleep, unable to concentrate their attention in productive, constructive ways.

GABA is the predominant neurotransmitter in the human body; there is a thousand times more of it available than all the other neurotransmitters combined and GABA is present at 40% of all *synapses,* or connections between nerves. It is inhibitory, serving to reduce the number of neurons firing, impeding neurotransmission, and thus calming the nervous system down from an excited state. GABA performs this function by binding to receptor sites on a *ligand-gated* neuron, which causes it to open a channel, allowing an electrically charged particle of chloride to enter. A *ligand* is an ion, a molecule, or

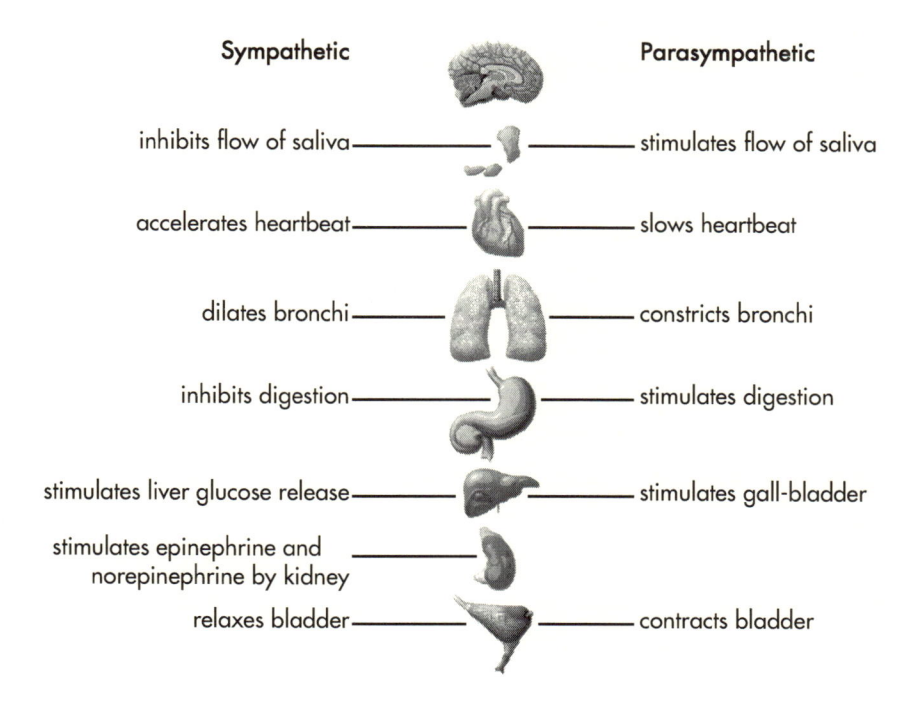

Sympathetic		Parasympathetic
inhibits flow of saliva		stimulates flow of saliva
accelerates heartbeat		slows heartbeat
dilates bronchi		constricts bronchi
inhibits digestion		stimulates digestion
stimulates liver glucose release		stimulates gall-bladder
stimulates epinephrine and norepinephrine by kidney		
relaxes bladder		contracts bladder

FIG. 3 : *Autonomic Nervous System*

even a group of molecules, which binds to a molecular structure. Thus, it is the binding of the ligand, GABA, that causes the neuron's 'gate' to open to allow entry by chloride. This action maintains the electrical potential of the cell, making it less apt to 'fire' electrically and pass its information along the other components of the nerve complex. Our subjective perception of what we feel like when nerve cells are less active is one of calmness, so when GABA performs this function, it promotes our sense of being calm or relaxed.

Benzodiazepine binds to a subunit of the GABA receptor called the 'BZD receptor.' It then 'potentiates' the action of GABA, increasing its ability to open the ion channel. Substances such as alcohol, opiates or barbiturates can actually *cause* the chloride channels in nerve cells to open but benzodiazepine, however, does not. It only enhances the action of the natural chemical which performs this function, GABA.

In the same way that the positive pole of one magnet will attract the negative pole of another, GABA floating freely is attracted to the GABA receptor sites where it 'binds' or sticks. Because benzodiazepine makes GABA more effective, less GABA needs to be attracted to the neurons while still performing the function of maintaining calmness. The 'affinity' the GABA receptor sites have for attracting GABA is reduced because of the presence of benzodiazepine.

Unfortunately, even after benzodiazepine use has been discontinued and there is no more of the chemical in the body, the GABA-α receptors in some individuals retain their diminished affinity for attracting GABA to the sites. The result of that lowered affinity is that insufficient amounts of GABA are utilized in mediating the nerve cells' impulses, and this can have far-reaching effects since so many bodily functions are influenced or controlled by the nervous system.

We tend to think of benzodiazepines only in terms of the most dramatic effects they produce: calming anxiety, overcoming insomnia, stopping muscle spasms. As such, the primary withdrawal symptoms people experience when discontinuing benzodiazepines are an increase in anxiety levels and difficulty sleeping, whether an inability to fall asleep or to stay asleep for a sufficient amount of time. In fact, as the body becomes habituated to benzodiazepine, 'rebound anxiety' or 'rebound insomnia' may occur, where the original problem with anxiety or insomnia may come back with greater intensity than it had when the drug was prescribed. Rebound effects often occur upon discontinuation since benzodiazepine is no longer potentiating the GABA that promotes calmness or sleep, and may last for a considerable length of time. A brief analysis of the etiology of rebound anxiety yields an immediate insight to the benzo problem: the human body's own anxiolytic agent is GABA. Benzodiazepine increases anxiolysis by potentiating GABA. If, after discontinuation, the body has greater anxiety than before benzodiazepine use began, the logical explanation for that to happen would be if the use of benzodiazepine had altered either the amount of available GABA or the ability of neurons to utilize GABA.

Many of the other symptoms of withdrawal are therefore entirely 'rebound' effects, in that they were not present before benzodiazepine was originally taken. Fasciculations, muscle spasms, and other symptoms which only appear after benzodiazepine has been reduced or eliminated from the body are direct results of the down-regulation of GABA mediation of the nerve cells. Because the Central Nervous System, the Sympathetic Nervous System, and the Parasympathetic Nervous System are all comprised of cells which depend upon GABA to maintain their equipoise between states of excitability and calmness, any of the bodily functions those systems control may be affected by benzodiazepine discontinuation. For example, the Parasympathetic Nervous System releases the chemical acetylcholine, stimulating the parotid, submandibular and sublingual glands which produce saliva. Without sufficient GABA to inhibit its activity, the Parasympathetic Nervous System may trigger the release of too much saliva, something commonly reported in people who have difficulties coming off of benzodiazepine. Problems with digestion or elimination of food may result from benzodiazepine discontinuation because GABA is present in tissues in the gastrointestinal tract. The 'feel good' neurotransmitter, serotonin, is known to interact with certain GABA receptor sites, so an abnormality in affinity at these sites may cause imbalances between GABA and serotonin that might explain why *anhedonia,* the inability to experience pleasure, and *dysphoria,* a pervasive negative mental state, are common sequellæ of benzodiazepine discontinuation in some people. It is certain that investigation of almost any of the myriad withdrawal symptoms commonly reported would reveal as its source a malfunction in the nervous system due to insufficient GABA.

People who go into rehabilitation to get over alcohol or opiate addictions also struggle with rebound anxiety and rebound insomnia, but such problems subside in a relatively short time, often in a matter of mere days after completing the discontinuation process. **While the down-regulation of GABA mediation of nerve cells after long exposure to benzodiazepine is commensurate with that of other substances, the potential for severity, diversity and persistence of withdrawal symptoms distinguishes benzodiazepine recovery from most others.** With this in mind, it would be far more accurate to say that people who are dependent upon benzodiazepine are not so much addicted to the benzodiazepine itself, but to GABA, and their suffering is a result not of a lack of benzodiazepine but a lack of their own bodies' GABA, which benzodiazepine has caused to be unavailable in sufficient amounts.

Chapter Ten
Supplements

Since the source of benzodiazepine discontinuation syndrome is known, is there any remedy for it? The short answer is, no. Except for the passage of time, there is nothing known that will cause neural receptor sites to up-regulate their affinity for attracting GABA. Normally, the body itself heals infirmities, but there is little to trigger its healing response. 'Healing' is not even an appropriate word, as it implies a reaction to disease or tissue damage. In this problem, there is neither condition. The part of the body affected is a microscopic area of the nerve cells, the tiny places where GABA binds, and the tissue there is not damaged or diseased; it isn't even non-functional, which might provoke a response by the body to replace the tissue. It is simply *less* functional than it is at its optimal level of operation so the body does not recognize it as something requiring healing.

A cure for the condition would be a substance or remedy that would reach these miniscule points on the nerve cells and affect them in such a way as to restore their capacity for attracting GABA. As there is no such substance or remedy known, the next question would be, is there anything that would at least relieve the symptoms brought on by this condition? The short answer there, too, appears to be, no.

It can be reasoned that since there is an insufficiency of GABA being utilized, it would help to increase GABA by taking it dietetically. There are GABA supplements available, but GABA doesn't pass the blood/brain barrier so ingesting it would not cause it to reach the brain. More importantly, there is nothing to suggest that the amount of GABA available in the body is diminished; rather, it is the ability to attract it that is diminished. One variant on dietary GABA is a preparation developed in Russia called 'picamilon.' It is comprised of GABA and niacin bound together in a single molecule. Since niacin readily transverses the blood/brain barrier, it is thought that this method can convey GABA directly into the brain. Extensively tested for toxicity and efficacy in Russia, picamilon has yet to be studied much in the West so there isn't much reputable science about it, nor is any likely: almost all such scientific studies are performed only when there is a profit motive for pharmaceutical companies.

There are other dietary supplements thought to assist in GABA functions. Niacinamide, also known as nicotinamide, has been reported to potentiate GABA in a manner similar to benzodiazepine, but without binding to the GABA-α receptor sites. Theanine, an extract from the tea plant, increases the formation of GABA. High levels of taurine have been shown both to activate GABA receptors and to increase the amount of GABA in the brain. An amino acid synthesized in the liver from other amino acids, taurine is also available as a dietary supplement, both by itself as well as bound to other nutrients, such as magnesium taurate.

There are pharmaceutical drugs which also target increased efficacy of GABA. Pfizer markets both *Neurontin* (gabapentin) and a more potent drug, *Lyrica* (pregabalin) as treatment for nerve pain and as an adjunct therapy for partial onset seizures in adults. Although the action of these drugs is unknown, they have been prescribed for off-label use by physicians to treat depression, anxiety, and related problems as GABA deficiencies. This is probably because both pregabalin and gabapentin are remarkably similar to GABA, and as such, are known as 'GABA-analogues.' They are not, however, 'GABA-mimetic,' *i.e.,* they do not function in the same way GABA does, and their results are, therefore, different than those caused by natural GABA actions. For this reason, pregabalin and gabapentin cause side effects not associated with GABA. One peculiarity of gabapentin is that as the dosage increases, the bioavailability of the drug decreases—the greater the dose, the lower the percentage of the drug used by the body.

There are even 'nutriceuticals' available as GABA modulators. *Nutriceuticals* are formulations of nutritional substances combined with the intent to provide specific, targeted therapeutic actions in the same way pharmaceuticals are meant to do. One such product, *NeuRecoverBZ,* is marketed specifically for people who are suffering as a result of benzodiazepine discontinuation. It provides a proprietary blend of vitamins, minerals and amino acids thought to support recovery and to mitigate the withdrawal associated with coming off benzos. Their marketing effort includes one single 'study' of thirty-five men and women experiencing acute and protracted withdrawal symptoms after having reduced their intake of benzodiazepine. All but two of the volunteers participating in the study reported some benefit from taking NeuRecoverBZ. The study was not conducted with scientific formality, however, and the results could hardly be viewed as conclusive in any way.

The paucity of competent studies means that information about what may or may not help is most often testimonial and, therefore, highly subjective. In the years I spent as an active participant in on-line discussions about benzodiazepine recovery, many people reported taking the substances listed above, and others as well. While a few people may have felt a modicum of relief, the preponderance of anecdotal evidence about the efficacy of supplements was simply that they did not provide enough benefits to justify their cost. In a population of traumatized people, desperate to find something to alleviate their suffering, anything truly beneficial would be welcomed as a godsend. If anything, such people would be predisposed favorably rather than negatively when assessing the effectiveness of supplemental therapies.

Since antiquity one of the hallmarks of civilized peoples has been their search for remedies to treat their ailments; so ingrained is this trait by now that it can almost be regarded as human nature. In my more desperate moments while tapering with Valium, I tried NeuRecoverBZ, niacin, niacinamide, magnesium, calcium, 5htp, Sam-E, passionflower extract, theanine, valerian root extract, and melatonin. I tried *picamilon,* that product made in Russia which combined GABA and niacin in one molecule. Since niacin readily crosses the blood/brain barrier, reports seemed to confirm that picamilon was an effective method of increasing GABA. Not only were picamilon and the rest of these supplements of no noticeable benefit, some produced quite negative results. I took supplemental melatonin to promote sleep, but after two days, found myself growing increasingly depressed. The depression lifted upon discontinuation of melatonin. NeuRecoverBZ had a paradoxical effect upon me, increasing my anxiety and overall feeling of malaise.

In fact, many common supplements almost universally regarded as beneficial to bodily health appear to cause paradoxical reactions in people who have difficulties due to benzodiazepine discontinuation. The B vitamins, helpful in combating the effects of stress, tend to cause nervous excitation characterized by anxiety. Fish oil, with its beneficial omega-3 fatty acids, most often provokes a similar paradoxical response. The only preparations that appear to be well tolerated and provide a mild modicum of relief are a mixture of lemon juice with water, and occasional use of Rescue Remedy, a tincture of medicinal flowers with a small amount of alcohol, useful in quelling anxiety.

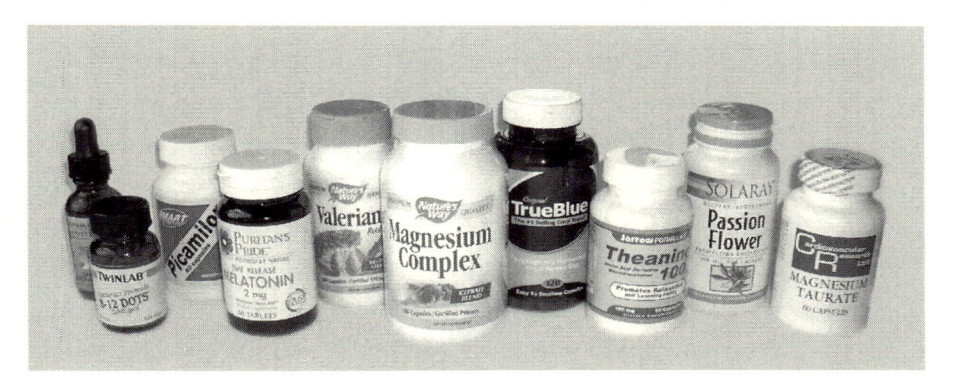

FIG. 4 : *Typical Supplements*

On the evidence, the only true help for recovery is the passage of time, while neurons throughout the nervous system restore their ability to attract GABA. All that can be done to facilitate the process is to minimize any excitatory stimulation which would retard it. The time frame during which recovery occurs varies with each individual. In some, restoration is so quick that no benzodiazepine withdrawal syndrome ever appears, leading to the perception by most members of the medical establishment that such a condition doesn't exist. In others the process takes months or years after discontinuation of benzodiazepine is completed. And for an unfortunate few, recovery is remarkably lengthy, a condition known as *Protracted Withdrawal Syndrome*. To such people, the passage of time is an agonizing experience.

As dire as this sounds, however, the time does pass and neurons do begin to regain their ability to attract GABA. By using the methods of Professor Ashton, *i.e.,* by tapering slowly and by reducing by no more than a small percentage of the overall dose with each cut, the restoration of normal GABA function may begin even while still taking benzodiazepine.

In my own experience, the early days of my tapering with Valium were marked by constant anxiety and by the near constant flow of adrenaline. As I have already mentioned, I owned a powerful sports car with a V-8 engine, which I could barely drive. If I stepped on the accelerator pedal with the slightest amount of excess pressure, the engine would roar and the rear tires would spin. The adrenaline coursing through my system would increase, my hands would tremble, and I would go into a mild state of shock. After a few months of tapering, however, I observed that the adrenaline had subsided considerably. I began to test myself by driving my car at an accelerated rate of speed. Sure enough, my anxiety level rose, which was appropriate since fast driving is a potentially life-threatening activity, and adrenaline increased. But after slowing back down to a safe speed, the anxiety abated, as did the adrenaline flow. The only way that could happen was if GABA in my body was being attracted in sufficient amounts to calm the excitatory state I had invoked. That my nervous system could calm itself down even while reducing the amount of available benzodiazepine in my body was proof that the process was working.

Another example of this was the fact that, at any given time during the taper, had I suddenly stopped my benzodiazepine intake, I would have gone into a horrendous cold-turkey reaction. Without the presence of the drug at the GABA-α receptors, I would not have been able to function. And yet, with slow, step-wise reductions in the dosage, I was able to maintain a relatively functional state. This was evidence that the neural receptors were *increasing* their ability to attract GABA, even as the amount of benzodiazepine present at the receptor sites was diminishing.

The process by which recovery occurs can be, arguably, maddeningly slow. But I was grateful that at least there was such a process at all. There are so many physical conditions from which there is no recovery. Thankfully, Professor Ashton's research and clinical work had provided a safe method for accomplishing the daunting task of getting off of benzodiazepines, while minimizing the distress as much as possible.

Chapter Eleven
The Emotions of Survival

The single overarching withdrawal symptom I experienced could be characterized as inappropriate anxiety, a deep and systemic neural stress that manifested itself in innumerable ways and persisted long after I had discontinued taking Valium. I view it as 'inappropriate' anxiety because it arose out of neurochemical conditions within the body, rather than as a response to authentic external stimuli. As evolved modern humans, we tend to think of anxiety as a negative, a state to be avoided; as such it is often medicated away. In reality, anxiety is a useful emotion.

Our early reptilian ancestors had quite rudimentary brains, often merely a cluster of nerve cells at one—or sometimes both—ends of the spinal column. Such primitive brains were capable of primitive consciousness, and primitive emotions as well, centered entirely on the biological imperatives. Each animal's personal survival depended upon finding food and water, resting when fatigued, avoiding or evading predators. Survival of the species depended upon procreation. These needs gave rise to corresponding emotions. Without hunger, an animal may not have been motivated sufficiently to locate food, and lack of sustenance would then have caused it to grow physically weak, a condition threatening to its survival. The opposite emotion, satiety, would tell the animal that it had eaten enough. Without satiety, it might eat so much food as to overload its digestive system, slowing it down and thus making it more vulnerable to predators. The urge to breed was, perhaps, too abstract and long-term a concept for creatures with limited intellectual and emotional capacity, so reproduction was integrated with pleasure. The urge to breed was experienced as the urge for sexual gratification, or lust. Even 'tiredness' is an emotion, although it is not recognized as such; we say we *feel* tired or we *feel* sleepy, and the act of getting into bed to go to sleep is usually perceived as a feeling of relief.

Rage was an important emotion to carnivores. It takes a great deal of energy to hunt down another animal and then do enough physical harm to it to cause its death. Rage provided an emotional stimulus beyond hunger to compel an animal to expend enough effort to kill. The opposite of rage, in the context of survival, is fear. When faced with the sudden threat of imminent death, fear can cause a potential victim to exhibit extraordinary physical abilities in running away from a predator.

While fear is the response to an immediate and definite threat, anxiety is a more generalized fear arising from a perception that something in the environment *might* pose a threat to survival. To prevent it from being surprised by a predator, anxiety elevates an animal's level of caution, maintaining its vigilance over a period of time so that it is constantly prepared to invoke the more dynamic fear response if necessary. On the predator's side of the equation, anger is the generalized form of rage. Perhaps perceived as a vague anger at feeling hungry, this emotion would stimulate a predator, keeping it focused and alert as it searched its environment for potential prey.

All of these emotions are transacted in the body by means of neurochemical processes. Their expression is accomplished by the action of different neurotransmitters, *i.e.,* the amino acids, peptides, monamines and acetylcholine, which cause physiological responses. The axis between rage and fear is the emotional aspect of the 'fight/flight' response critical to the survival of our ancient forbears, and its functional components remain intact in the human body of today. As mammalian—and then human—bodies evolved past their ancestors, their more complex brains did not replace the reptilian brain stem; rather, their structures were overlaid on top of it. The base emotions of our 'earliest brain,' however, are generally further refined by the influences of our mammalian, or limbic, cortex, and more importantly, by our neocortex, the 'thinking' brain whose high degree of development differentiates humans from other species.

Since regulation of these complex systems depends upon the balanced employment of neurotransmitters, it is easy to recognize that a malfunction with GABA, the most prevalent of neurotransmitters, would have a disruptive impact on the effective use of all the others. GABA's primary function is *anxiolysis,* the reduction of anxiety, tension or agitation. Failure of GABA to modulate excitability therefore may result in self-generating excitatory states in the nervous system. Most often, these reflect the 'flight' axis of the fight/flight stress response, but in some individuals, the 'fight' response is invoked instead and irrational outbursts of rage occur during benzodiazepine withdrawal.

Such outbursts and other phenomena have been reported during regular benzodiazepine use as well. The artificial enhancement of GABA from having benzodiazepine bound to GABA-α receptor sites can cause disinhibition, where normal controls fail to suppress primal impulses. Any of the emotions native to the reptilian cortex can emerge unrestrained, sometimes leading to violent behavior when inappropriate rage occurs, or sexual misconduct if inappropriate lust should arise. Such circumstances can occur during benzodiazepine discontinuation, but the great preponderance of symptoms comprises manifestations of the emotions of the flight response: anxiety or fear.

The stress response, *i.e.,* 'fight or flight' is intended to ready the cardiovascular and musculoskeletal systems to deal with a life-threatening event. Almost by definition, such an event would be a brief one: an animal would either survive the fight (or not), or successfully evade a pursuer (or not.) In either case, the event itself would be resolved in a relatively short amount of time. With a self-generating stress response, however, one that appears not as a reaction to external conditions but because of internal, neurochemical ones, there is no signal that will turn off the process. Anxiety and fear may then become dominant characteristic emotional states causing a near constant state of flight—even though there is nothing to flee from.

Since the purpose of anxiety is, at its root, to maintain survival, its emotional messages are somewhat imperative, tending to override everything else. To an animal in a perilous natural environment, anxiety fosters a heightened awareness of what is potentially wrong, rather than what is right or positive. Any feelings of contentment would tend to contradict the idea that there may be danger present, so anxiety dispels such feelings as they may threaten the animal's survival. Therefore, the effect of anxiety is that it renders almost impossible the ability to feel contentment in any of its forms: feeling at peace, feeling at rest, feeling optimistic, feeling happy. While that may be valuable in the short term as a reaction to possibly dangerous circumstances, when anxiety is generated in the nervous system itself and continues unabated for month after month, the effect on the quality of life is horrific.

As I have related, my own method of dealing with anxiety was to assess it rationally and recognize that it was 'unreal,' *i.e.,* not happening because there was any genuine threat to my survival but because of an unfortunate problem I was having with my body's neurochemistry. Numerous times throughout the day, every day, I would remind myself of this in moments of 'self-talk,' in effect, training myself not to believe in my perceptions, which I knew to be distorted. The result of this practice was that I became able to function *in spite of* the anxiety I was feeling and, in time, I came to ignore the sensations associated with it, the butterflies-in-the-stomach, jittery agitation, racing heart, even the continuous spikes of adrenaline.

I applied the same method to other withdrawal symptoms as they emerged. When the muscles in my legs would twitch for hour after hour, I would tell myself, "This is a result of my having been exposed to benzodiazepine. It's disturbing to experience, but it won't kill me or cause permanent harm. It has no meaning, no significance whatsoever." By using cognitive skills to mediate the experiences I was having, I was able to view many of them as little more than annoyances.

Chapter Twelve
"Don't take it personally"

More than anything else I did, it was developing the ability to look somewhat objectively at what was happening to me and to ground myself as much as possible in what I knew to be reality that helped me to cope with the devastation I was experiencing. I knew that my perceptions were unreliable, influenced as they were by a neurochemical perversion of natural processes, so I learned to believe not in what I felt, but in what I thought.

It is human of us to endow our experiences with meaning and significance. We quite naturally identify ourselves with what happens to us in our lives. And yet, in the case of devastation from benzodiazepine withdrawal, that's hardly an appropriate thing to do. The devastation didn't happen as a result of making a bad choice or an immoral choice, but merely an unfortunate one. Had I chosen to use heroin or cocaine, knowing it was a dangerous drug but thinking I could get away with it unscathed, there would have been a moralistic cause-and-effect relationship to my subsequent addiction problems. But that wasn't the case at all. I took a drug my doctor had assured me was safe, and I simply believed him. He was most probably unaware that long-term benzo use can present remarkable problems for some people—I'm sure that to this day, even though he 'supervised' my taper, he still has no idea of the hell I went through.

My doctor had put me on Effexor and Xanax originally because I had had an episode of severe anxiety disorder. My wife had almost died of pancreatitis, and after she got home from the hospital, I had what used to be called a 'nervous breakdown.' My doctor prescribed Prozac, and when that didn't work, he substituted Effexor. The Effexor caused 'agitation,' a common side effect, so he prescribed Xanax for the agitation. A couple of months after I had been taking Xanax I mentioned to a therapist I was seeing that I had begun to reduce my dosage in order to discontinue it. She asked me why I was doing that. I responded by saying that Xanax was a tranquilizer and I did not want to risk getting addicted to it.

She thought for a moment, then said, "I don't think there's much likelihood of that happening here. And you're doing so well these days. I think you should stay on the Xanax."

In the early days of my tapering with Valium, I often went back to that pivotal moment. If only I hadn't mentioned that I was discontinuing Xanax to my therapist! If only she hadn't advised me to keep taking it, I might have avoided this living nightmare! But I didn't blame her for my addiction, nor did I blame my doctor. My therapist was a caring professional. She wasn't making any money from my Xanax habit and had no venal interest in promoting or supporting it.

She simply was ignorant of the fact that she was putting me in danger by suggesting that I stay on Xanax. My doctor operated under the same ignorance. When I told him that I was experiencing certain unpleasant side-effects from Effexor, he listened to my description of those side-effects, diagnosed them as 'agitation', then prescribed Xanax to counter the agitation because that is what the protocols for Effexor use suggested. These protocols were approved by our nation's Food and Drug Administration, so my doctor simply trusted in their safety and efficacy, and I, in turn, simply trusted my doctor and my therapist.

As such, there was no malevolent significance to my becoming an iatrogenic drug addict. I simply happened to be one of the people whose GABA-α receptors fail to recover quickly their affinity for attracting GABA. But just because I wasn't angry with my doctor and therapist does not mean I was not angry. Once it became clear to me that my reaction to a supposedly safe drug had resulted in my becoming mentally unhinged and physically disabled, as well as addicted to a substance I could not stop taking without the danger of worsening my condition exponentially, I was furious. I felt absolutely betrayed.

I also felt appalled at the situation I found myself in. My mental abilities were compromised, but I had to manage carefully my daily dosage of Valium, a feat requiring a degree of focus that was profoundly difficult for me. I was concerned that if I got into trouble, there was no place where I could find help. I knew from reading the tragic accounts of others at *benzo.org.uk* that if I went to a mental hospital I would only be given more drugs, further deepening my predicament.

My resentment built until one day it reached a climax. "Life's not a vacation, you know," I heard myself think. "Something happens to everybody—this is what happened to you. Some people win the lottery, others contract cancer or get hit by a bus. This addiction to benzos is what happened to you. You're going to have to accept that. Being pissed off at the world is a waste of energy and it's not helping you, so get over it."

My addiction didn't happen because I was a bad person; God wasn't punishing me. No one had done this to me on purpose. It was simply a tragic mistake and I had to make the best of it.

Fortunately, thanks to Professor Ashton, I knew what the problem was, *i.e.,* down-regulation of the GABA-α receptors, and I had her method for safe discontinuation. To follow her protocol meant that I had *years* of a systematic course of action ahead of me, so I resolved to go about it as soberly and as dispassionately as I could. That is not easy to do, when the nature of what was physically wrong with me also affected me mentally and emotionally. 'Benzo hell' turns out to be an accurate description of what the ordeal feels like subjectively.

It was easy to conclude that one thing that would slow the recovery process considerably was stress. Stress would lead to trauma, and trauma would further burden an already-compromised nervous system. Often, that sets up a circular syndrome, where the stress causes trauma, and the trauma causes more stress, which further perpetuates the trauma. I realized that one of the most important things I could do to protect myself against needless additional stress would be an objective mindset. I knew I had a malfunction at a deep level of the basic operation of the human system. That's truly all it was. Subjectively, however, this condition often left me feeling sick, crazy, fearful, hopeless. It took great discipline to be in the middle of such feelings and tell myself, "I am *not* actually 'sick,' no matter how this looks or feels to me or others. My brain simply needs to repair a malfunction, which it will do on its own if given the chance. My job is to give it that chance, and not to take all of these things I'm experiencing personally."

Given time (and minimizing other stressors wherever possible) I knew I would heal. Why? Because, simply, the body is a self-healing mechanism. Encoded throughout all of its components are the instructions for repairing itself. As an example, if we cut ourselves, blood will automatically clot, the cut will close itself up, new tissue will form, and the problem will be repaired—all without our conscious control. What causes a cut not to heal is *interference* with the healing process: not keeping it clean, picking at it, irritating it, stressing it out by expecting (or demanding) that the damaged area perform while it's still recuperating.

What had happened to my brain and central nervous system was far more grave than a simple cut. Recuperation would take much, much longer as a result—but the underlying principles were the same. It would heal itself if the interference with its healing were minimized, and the recuperation were supported in every way possible. Based upon that idea, I gave myself permission to avoid stress as much as possible, and to focus all of my attention on simply getting safely off of benzodiazepine.

While anxiety was the most dramatic emotional state I dealt with every day due to its physical symptoms, it was overlaid onto a foundation of profound systemic depression. I was simply incapable of experiencing any good feelings whatsoever, having any pleasant sensations or of thinking any happy thoughts. It is staggering to realize that the use of tranquilizers could result in such complete eradication of everything positive in a person's consciousness, but it's true; it happened to me and has happened to many others, as well. I recall standing in the bathroom, looking in the mirror, practicing how to smile. I had realized that I hadn't smiled in half a year or more and was trying to get myself to do it. My first attempts were pathetic, the corners of my mouth upturned mechanistically, while my eyes retained their feral sadness. Even when I was successful in imitating a true smile it did not bestow upon me any human warmth. The effect upon my psychology of such pervasive and perpetual negativity was distressing. When I was finally once again capable of an intellectual examination of what was happening to me, I realized I had to do something about my psychological state. I was having thoughts such as, "I am so depressed. Life is just an ugly, painful ordeal." But I knew that that wasn't what I really thought. I had always been at my core a cheerful, enthusiastic, optimistic person, often in spite of external circumstances. 'Life' itself hadn't changed; only my perception of it had changed.

By looking at what was happening to me with dispassionate objectivity, I could see that my thinking, "I am depressed" was, if nothing else, inaccurate. It would have been more nearly correct—and less damaging—to say, "I am experiencing depression right now" because in saying, "I am depressed" I was defining myself as a person based upon symptomatic phenomena. Depression, mental confusion, exhaustion, these things and many more are symptoms of benzodiazepine drug usage or discontinuation. To define myself based upon chemically induced subjective experiences was not in the slightest bit an accurate or realistic thing to do. Saying, "I *am* depressed" was a form of self-programming: I realized that if I said, "I am depressed" enough times, depression would then become the lens through which I viewed absolutely everything. It would be far healthier, and more true an assessment, if a little long-winded, to say, "I am experiencing depression as a result of imbalances in my brain chemistry, brought on by using an unpredictable drug which I was led to believe was safe. In time, my brain will reestablish the delicate balances between the chemicals it produces and uses, because my body is a naturally self-healing system."

As much as my often bizarre and sometimes horrifying symptoms demanded my attention, I understood that the GABA problem was the single cause of all my distress. In the grip of anxiety, panic, fear, doubt, hopelessness, brain-shivers, and more, it was almost impossible to keep that fact in mind, but I knew it to be the biochemical truth. The extent of my difficulties lay in that one problem: temporary loss of affinity to the naturally-occuring brain chemical, GABA. Feeling depressed and hopeless, I had been projecting my hopelessness onto my subjective life experiences, and drawing irrational inferences from them—because I was discounting the objective reality that the source of my distress was a problem with my nervous system, not my life itself.

Having the condition of dysphoria, my thoughts, quite understandably, were unhappy, depressive thoughts. Thoughts by themselves are benign, but when we 'personalize' thoughts, we then identify ourselves with them, determining our identity (who we believe we are) by the thoughts we think. That appears to lead to a vicious cycle mechanism in that this person we believe we are would think thoughts such as these and therefore does. The thoughts reinforce the belief—and a 'persona' gets established.

This is tricky stuff indeed for anyone, but for people like me, in benzodiazepine use and/or withdrawal, there are additional chemical influences that affect our thinking. If we then identify with that thinking, we're in effect basing our beliefs about who we are on malfunctioning neurochemical processes. This isn't our 'authentic' selves, it's the result of biochemistry gone awry. To make matters worse, we're also often mentally confused anyway so we end up making determinations about who we are and what life is, when we're really in no condition to make them.

The insight that the symptoms I was experiencing were clearly due to the effect of a drug formed the basis of my 'objective' approach to how I chose to perceive what was happening to me. In other words, we tend to think that the things we experience are personal to us, and reflect our reality. That was clearly not the case for me while I was discontinuing benzos: the things I was experiencing were a result of the effect of drugs upon my brain/central nervous system, which in turn was affecting me mentally. Armed with that knowledge, when I was experiencing depression, for example, I was able to avoid making the mistake of defining myself by saying, "I am depressed" or "I am anxious" when I knew that the source of the depression or anxiety was neurochemical, not anything in my personal reality. Being able to think, to remain objective in spite of the horrendous negative emotions and experiences that benzodiazepine was causing me to have was the one thing that kept me relatively sane. The use of cognition, of *thinking*, was perhaps my greatest tool for getting through the ordeal.

Chapter Thirteen
Persistent Depression

Over the course of the next year, I had a total of five days where I actually felt okay. One occurred in springtime— out of the blue, I had a day where I felt normal. The other four happened the following late August and early September. One remarkable feature of those days when I felt okay was that I almost instantly forgot what it was like to live in horror. I simply was back in the groove I was always in. I noticed it at the time, and then observed that the same phenomenon happened in reverse when the flip side occurred and I re-entered depression. I recognized that when I felt bad, I almost couldn't imagine what it was like to feel good, or that I would ever feel good again. I felt like I was trapped inside myself, feeling terrible, that there was no escape for me, no place I could go for comfort or where I would feel better at all.

On those rare occasions when I did feel okay, I characterized it as feeling 'like myself.' Of course, the obverse of that was that when depression was at work, it was easy then to believe that feeling depressed was 'like myself.' It took continual cognitive effort not to succumb to that idea. Interestingly, feeling 'like myself' when I was not depressed somehow seemed to be my authentic perception, a 'real' perception, where those inspired by depression were somehow 'false.'

It was important to me to remember—and really to keep firmly in mind—that feeling good is actually normal. 'Feeling good' simply means that the body is consuming oxygen (which is why exercise and movement 'feel good') and that it is producing and using serotonin, norepinephrine, dopamine, GABA, and the other essential neurotransmitters, amino acids, hormones and vitamins effectively. When the antidepressant drug, Effexor, had lifted me out of depression ten years earlier, it had done so by augmenting how my body utilized serotonin and norepinephrine. That was the extent of its 'magic.' It didn't change reality—it couldn't. It was just a chemical. But it could change how my body influenced my perception of reality, and during the period when it was effective, the change was for the better.

I drew some conclusions about the fact that when I started to feel good instead of depressed, I didn't feel like a 'different person,' I felt 'like myself.' My normal state is evidently to be happy and productive, and I'm happy and productive when all of my physical systems work the way they were designed to work. Depression is an unnatural state to be in: according to the prevailing idea of psychiatry, it occurs when the body's neurochemistry is out of balance. Sometimes stress or the effect of traumatic life experiences can, according to this theory, create a chemical imbalance. It can also happen from an environmental toxin creeping into the system, from the internal 'body-clock' that regulates sleep not functioning correctly, as a sequella to certain disease states, from dietary deficiencies, or from a host of other causes, both known and unknown.

Depression can also result from the use of psychotropic medicines (psych meds), powerful synthetic chemical compounds that affect brain functions, the most common of which are Selective Serotonin Reuptake Inhibitor (SSRI) type antidepressants. For those who have had a positive response to SSRIs, depression may result if an attempt is made to discontinue their use. Antipsychotic drugs and neuroleptics are often prescribed for off-label use, such as to counter sedative-resistant insomnia. With their profound effect upon mental states, it is no wonder that discontinuation may bring about severe emotional conditions such as depression and even psychotic delusions. And, perhaps most frequently of all, doctors routinely prescribe benzodiazepines, which directly influence the GABA-α sites that affect not only the brain but the entire central nervous system. After becoming accustomed to psych meds modulating the action of brain functions, neurological systems cannot help but be perturbed in the absence of those chemical drugs during discontinuation.

But since the brain and central nervous system are the instruments that we use to have a sense of who we are and what life is for us, when they are compromised by whatever cause, we feel terrible and draw the conclusion that we are unhappy people and that our lives are miserable. The only way we can hope to understand that those are erroneous conclusions is by thinking—a lot of thinking, a lot of repeating the same thoughts, over and over again, to keep us from believing the messages emanating from our damaged senses, and to help us maintain more realistic expectations; mainly, that once our bodies have reestablished their natural chemical balances, the bad feelings will disappear and be replaced by feeling good, our normal, healthy state.

Chapter Fourteen
The Chemical Imbalance Theory

The understanding that depression and other mental conditions are caused by a chemical imbalance in the brain is now well established, and serves as the basis for how psychiatric therapy is now practiced. Where once psychiatrists would listen to their patients as they recounted their dreams or memories of childhood for fifty-five minutes, they now prescribe drugs; a session with a psychiatrist is a brief affair, the result of which is either a renewal of the current prescription, a change in dosage, or a change in medications. Television advertisements for Zoloft have stated:

While the cause is unknown, depression may be related to an imbalance of natural chemicals between nerve cells in the brain. Prescription Zoloft works to correct this imbalance. You just shouldn't have to feel this way anymore.

Other direct-to-consumer-advertisements by Zoloft's competitors make similar claims. Medical consumers feel they can place a certain amount of trust in the television commercials and print ads for antidepressants because they know that regulatory agencies, such as the Food and Drug Administration in the United States, do not permit advertisements that are not supported by scientific evidence.

The theory that mood disorders have physiological—rather than psychological—origins has removed much of the social stigma of having such a disorder, and that has led greater numbers of people to seek 'proper medical treatment' from their doctors. Direct-to-consumer advertising results in people asking for specific drugs by name, Prozac or Celexa for depression, Paxil or Effexor for Social Anxiety Disorder, for example. Studies have shown that physicians are statistically more likely to prescribe medications patients have requested by name than to refuse them. In the April 27, 2005 issue of *The Journal of the American Medical Association*, a study by Richard L. Kravitz, MD, MSPH, director of the Center for Health Services Research in Primary Care at the University of California, Davis, found that, among other results, people who ask doctors for antidepressant drugs are much more likely to get them, even if their condition is mild and wouldn't be likely to respond to drug therapy; and that patients asking for a specific brand-name drug were far more likely to get that drug than another antidepressant. One obvious conclusion that can be drawn from Dr. Kravitz's study is that the $3.2 billion dollars pharmaceutical companies invest in drug ads each year is well spent: the advertisements are highly effective in eliciting specific responses from consumers—and their doctors.

While the belief that depression is caused by a chemical imbalance is now so widespread that it is regarded as fact, the reality is that there is no scientific evidence that would constitute proof of the theory.

The medical establishment suggests that mood disorders result from deficiencies (or overabundances) in neurotransmitters just as diabetes results from a deficiency in insulin production, and that both conditions may be treated by similar methods: the application of medicines to correct the deficiency. In the case of diabetes, however, the condition can be diagnosed by scientific tests of biological functions within the body.

Mental disorders, by contrast, are diagnosed by doctors using 'algorithmic' methods in which patients are questioned about their symptoms and, based upon their answers, assumptions are made about what disorders they must probably have. Such a subjective approach relies heavily upon the doctors' judgment in interpreting the information, a not-infallible process. Reported in the *American Review of Medicine,* Dr. Erwin Koranyi conducted a study of 2,090 psychiatric patients which revealed that 43% of them had undiagnosed physical illnesses that had been misdiagnosed as mental disorders. A study conducted in a Florida psychiatric hospital in the 1980s of one hundred patients believed to be mentally ill showed that nearly half of their psychiatric disorders resulted from underlying medical problems that had gone undetected.

In his book, *Prozac Backlash,* Dr. Joseph Glenmullen warns that, "[H]ypothetical biochemical imbalances have been presented to the public as established fact."

Yet, *there are no laboratory tests which show an overabundance or deficiency in a patient of any of the neurotransmitters believed to 'cause' mental disorders.* In the absence of such empirical evidence and the ability to employ the scientific method in determining a biological cause for such disorders, diagnosis is often made as a result of how a patient responds to medication. This can lead to the application of circular logic: since reduced levels of serotonin are believed to cause depression, if a patient is given a SSRI antidepressant and responds favorably to it, then a diagnosis of 'depression' can be made. Sometimes logic is abandoned altogether. Although serotonin is implicated in depression, the condition responds equally well to treatment by older tricyclic antidepressants, as well as by the more modern drugs, *buproprion* (Wellbutrin) and *reboxetine* (Edronax, Vestra), none of which works on serotonin. In fact, the mechanism of reboxetine is *exactly the opposite* of that of SSRI antidepressants, increasing the effect of norepinephrine rather than serotonin. The chemical imbalance theory also fails to account for the efficacy of talk therapy in relieving depression. How could a patient talking about their emotions significantly alter their own biology to improve a deficiency in serotonin? Such a method would hardly be suggested for a diabetic to improve the production of insulin.

In an article entitled, *Serotonin and Depression: A Disconnect Between the Advertisements and the Scientific Literature,* in the December 2005 issue of *PLoS Medicine,* Jeffrey R. Lacasse and Jonathan Leo state:

> *While neuroscience is a rapidly advancing field, to propose that researchers can objectively identify a 'chemical imbalance' at the molecular level is not compatible with the extant science. In fact, there is no scientifically established ideal 'chemical balance' of serotonin, let alone an identifiable pathological imbalance.*

They conclude by saying that,
> *The incongruence between the scientific literature and the claims made in FDA-regulated SSRI advertisements is remarkable, and possibly unparalleled.*

And this is quite certainly true. Because no scientific methodology supports the claims of SSRI antidepressant manufacturers that the drugs treat depression by restoring a 'chemical imbalance' to a healthy state, such claims are no more scientifically valid than those of witch doctors or snake oil salesmen for their wares. And yet, repeated calls upon the U.S. Food and Drug Administration to censor the pharmaceutical companies' false advertising go unanswered. One of the authors of the paper cited above, Jeffrey R. Lacasse, received this explanation in an e-mail from a regulatory reviewer at the FDA:

"Your concern regarding direct-to-consumer advertising raises an interesting issue regarding the validity of reductionistic statements. These statements are used in an attempt to describe the putative mechanisms of neurotransmitter action(s) to the fraction of the public that functions at no higher than a 6th grade reading level."

It seems unconscionable—and quite bizarre—that the U.S. Food and Drug Administration would consider it permissible for advertisers to present unverified, scientifically unsound information to sell products to people who may lack the ability to understand how they work. Karen Barth Menzies, an attorney for the Los Angeles/Washington, D.C. law firm, Baum Hedlund, which has represented several thousand SSRI victims in litigation, has said,

FDA has been violating its own mandate to act in the interests of the American consuming public by taking sides with the pharmaceutical companies it is supposed to police. The problem is not only the cover-up by the pharmaceutical industry, it is the FDA's lack of objectivity, which facilitates that cover-up. The consequences of this complicity has[sic], in far too many instances, led to tragedy and death.

The U.S. Food and Drug Administration appears to have been supportive of Prozac—perhaps improperly—from the drug's inception. In September of 2004, the Alliance for Human Research Protection (AHRP) sponsored a press conference at the FDA Public Hearing on Antidepressants and Suicide. Dr. Peter R. Breggin, a noted critic of medicinal psychiatry, presented a report in which he stated:

> *Prozac failed to demonstrate efficacy in its clinical trials. When this potential economic disaster for Eli Lilly and Company was discovered, the FDA offered a way out to the drug company. The FDA allowed the drug company to include in its efficacy data those patients who had been illegally treated with concomitant benzodiazepine tranquilizers in order to calm their overstimulation. With these patients included, statistical manipulations enabled the FDA to find the drug marginally approvable. Basically, Prozac was approved in combination with addictive benzodiazepines such as Ativan, Xanax, and Valium; but neither the FDA nor the drug company revealed this information.*

The suggestion that only a "fraction of the public" would fail to understand how SSRI medications work is inaccurate, since *no one* truly understands how they work. The relationship between such drugs and depression is not linear, not one of direct cause-and-effect as the chemical imbalance theory suggests. Perhaps the most reasoned statement that could be made about SSRIs is that, in some patients, selectively blocking the reuptake of serotonin has a beneficial effect upon depression in the same way that taking an antacid tablet has a beneficial effect upon heartburn. What is remarkable is that the pharmaceutical industry and its sales arm, the psychiatric industry, can operate under a false premise with the apparent complicity of the

U.S. Food and Drug Administration. By contrast, in 2002, the equivalent regulatory agency in Ireland, the Irish Medicines Board, barred GlaxoSmithKline from including in its patient information brochure their claim that paroxetine (Paxil) corrects a chemical imbalance.

Patients assume that because psychiatry presents itself as a science-based discipline, psychiatric drugs are being prescribed in a scientific manner. In fact, neither premise is true. This is not to say that psychiatry does not help people who suffer, as that is not the case. The manner in which psychiatrists offer their help reveals itself, however, to be rather slipshod, and would be deemed unacceptable in any other profession. For example, imagine if an auto mechanic responded to a request to fix a 'funny noise' by replacing one part after another, week after week—charging money each time—until the problem went away. That is, essentially, the methodology of prescribing psych meds. The psychiatrist offers first one SSRI antidepressant, then another and another until one seems to help. Next, the dosage is increased until an effective level is reached. If annoying side effects present themselves, additional drugs may be prescribed to quell them. After that, the psychiatrist sees the patient for a ten-minute appointment once per month to refill the prescriptions. Psych meds are, by definition, powerful in that they alter a person's perception of their mind. One ten-minute appointment per month hardly constitutes 'supervision' over someone who is, quite literally, tampering with his or her brain chemistry.

The ethical considerations of misinformation aside, if SSRI antidepressants provided everyone with the benefits suggested in pharmaceutical company advertisements, the issue of the chemical imbalance theory would be merely academic. The fact is, however, that while substantial numbers of people have reported having benefited from SSRI antidepressants, a significant percentage of those who take them experiences difficulties. In its first ten years on the market, Eli Lilly & Company's drug, Prozac, garnered over 40,000 adverse drug reports in the FDA's Spontaneous Reporting System, more than any other medication in the history of the program. Numerous lawsuits have been filed over SSRIs, in cases where people who took them committed murder, suicide, and other aberrant behaviors. With the potential for such reactions, the ten minutes per month of evaluation by a psychiatrist appears to be quite irresponsible.

Although I was in a state of depression characterized by almost continual misery, I was not tempted to tell my doctor about it. I knew that he would almost certainly have responded by writing me a prescription for an antidepressant, probably Effexor again. While that drug had been useful to me in the past, my neurology hadn't been compromised at the time I had first taken it, something my doctor never quite understood throughout the lengthy period of my tapering off benzodiazepine. Early on in the ordeal, when I thought what I had was some form of fatigue syndrome, I mentioned my almost complete exhaustion to the doctor. He responded by giving me a prescription for Provigil, a drug offered to people with sleep disorders such as sleep apnea and narcolepsy. Provigil is a stimulant, not something my central nervous system was in any condition to tolerate. I had a rough couple of days after trying it. Later, when I discussed my libido problems with the doctor, he wrote me a prescription for Levitra, a drug of the Viagra type. As I knew my difficulties had nothing to do with erectile dysfunction, I never even filled the prescription.

By this time, I understood well enough that there was something deeply wrong with the practice of medicine, particularly in the United States. With the rise of globalization and market-based economics, people with health problems had become 'consumers' rather than 'patients.' As consumers, we are the targets of marketers, and health issues make us a particularly vulnerable group. Unlike other products offered to us for sale, which we can assess on their merits and choose either to buy or not to buy, medical products are understandably particularly compelling because they offer us the possibility of relief from our ailments and whatever pain, discomfort or distress those ailments cause us. As consumers, what remains in our power is that we are able to give 'informed consent' as to whether or not we choose to avail ourselves of what is offered to us, but informed consent relies upon our being given all of the information relevant to a particular drug or procedure, and we medical consumers have little access to the sort of information which would lead to an informed choice.

Our regulatory agencies, which would seem to be the authoritative source of such information, appear to be more dedicated to the interests of pharmaceutical companies than to those of consumers, and thus don't protect us from the predations of marketers. In recent years, the Chief Counsel of the U.S. Food and Drug Administration has actively participated on the behalf of industry in private state litigation cases where people were suing pharmaceutical companies and the manufacturers of medical devices for damages their products caused. It seems unconscionable that government agencies in a democratic nation, putatively the representatives of the people, would be more favorable to industry than to the populace. It is perhaps believed that, in a market-based economic environment, anything that stimulates business is good for society as a whole. The expansion of the pharmaceutical industry may be viewed as having eventual trickle-down benefits to consumers. John Fetto, writing in the March 2003 issue of *American Demographics*, notes that in 2003, people in the United States spent $161 billion dollars on pharmaceutical products, four times as much as they spent in 1990, a figure expected by sales projections to reach $360 billion by 2010. It is conceivable that the people in government who influence the regulatory agencies see such dollar volumes as good for the nation and therefore encourage the role of the FDA as a facilitator to the pharmaceutical companies rather than as a watchdog protecting the health and safety of the citizenry. And yet, to achieve social benefits by taking money from the pockets of consumers for products of often questionable value seems antidemocratic.

Chapter Fifteen
Market-Driven Medicine

The arrival of Prozac in the medical marketplace, followed shortly by 'copycat' SSRI medications inspired by Prozac's commercial success, caused quite an impact on society. While Prozac was originally targeted to patients suffering from clinical depression, its user base quickly expanded to include anyone who could benefit by taking a pill that resulted in their 'simply feeling great.' At first, this seemed to be a win-win-win situation: consumers won, in that they experienced improvement in mood and ability to function in the stressful social environment modern life imposes; the prescribing doctors won by revenue generated from office visits which resulted in SSRI prescriptions; and most of all, the pharmaceutical companies won by providing a product that was so well-received by consumers.

Investments in research and development by drug companies yield numerous different types of medications which could be marketed to consumers. However, there is a commercially important difference between an antidepressant and one such as Tylenol. Tylenol is taken when someone has a headache, fever or similar complaint. When the problem has been resolved by the medication, its use is discontinued. Its benefits are understood to be temporary. An SSRI, on the other hand, is meant to be taken continuously over a long period of time. While SSRI medications are not cheap, their users will gladly pay the cost of taking them, and, in contrast to a headache remedy, they are pills that are taken every day, month after month, or year after year, effectively becoming an economic addiction. Unlike other products, an addictive product requires a marketing effort but once: to introduce it to individual consumers. Once they begin taking the product, they will continue to buy it, unquestioningly, almost in perpetuity— stopping only because it has become ineffective or has caused difficulties in the personal life of the user so intolerable that its use must be stopped. From a marketing standpoint, such drugs are nearly economically perfect: the product only has to be presented to a consumer once, after which the consumer will initiate any action necessary to insure the continued purchase of the product. Since the pharmaceutical company only has to 'sell' a drug one time, it then never has to use its marketing resources again in order to keep that consumer buying its product.

Further, the marketing of the product (the 'sales pitch') is delivered by a physician, a person trusted by the consumer to give recommendations. And uniquely, the consumer actually pays the salesperson (*i.e.*, the physician) for being introduced to the product. Once again, I don't mean to suggest, nor do I believe, that doctors are avaricious drug peddlers, interested only in making a profit by pushing the pharmaceutical companies' wares. A doctor's primary motive in his/her actions is to provide patients with relief from troubling, painful, or pathological symptoms. A patient presents to the medical doctor with such a symptom, and the medical doctor treats by prescribing medicine. The doctor may offer one medicine after another until the patient finally reports relief. At a psychological level, the doctor then feels a sense of 'reward', that he/she has helped someone in need. I believe it is the pursuit of such positive, life-affirming psychological rewards that drives doctors to practice their profession, rather than mere monetary incentives.

However, with the shift to market-driven medicines, the beneficence of doctors is perhaps exploited by pharmaceutical companies, and physicians become the unwitting vendors of pharmaceutical drugs. If the drugs were safe, this would not be a problem, other than a purely economic one. The drugs are not safe, however. And doctors have no way of knowing about the safety hazards of the drugs they push because the primary source of their information about them comes from representatives of the pharmaceutical companies themselves. The FDA cannot be trusted to supply accurate information about the potential harm from medicines because of the unhealthy relationship existing between governmental regulatory agencies, the pharmaceutical industry whose wares they are meant to monitor, and the insurance industry that facilitates the mass distribution of pharmaceuticals. Rather than having an effective system of checks and balances, the above-named collude to produce economic growth, though it be at the expense, literally and figuratively, of the public.

At the heart of these problems lies, not the drugs themselves, but rather the mindset which so aggressively promotes their use. This is market-driven medicine, rather than treatment-driven medicine, a phenomenon which began in the 1960's with the introduction of a cold remedy named *Contac*. Prior to the arrival of Contac, the leading medicine to provide relief for cold symptoms was Coricidin. As was the norm at the time, Coricidin was sold as a bottle of pills. Medicine chests held similar such bottles: one containing aspirin to treat headache and fever, one containing pills to treat occasional diarrhea, *etc.* Contac, however, was marketed—and packaged—differently. It was sold in a box, not a bottle, and each pill was isolated in a blister pack. By virtue of this packaging, each pill was perceived by the consumer to be of great benefit, relieving cold symptoms for up to twelve hours. And thus, twelve capsules could be sold for the same price that would formerly have been paid for a bottle containing perhaps thirty-six capsules.

While Contac may have introduced the concept, market-driven medicines have now reached a new level of sophistication, so new that there is no word to describe it and one will have to be coined. While *iatrogenic addiction* denotes an addiction caused by a doctor, *venditiogenic addiction* will describe a purchasing addiction caused by the seller of a product.

In its least complex form, the practice simply promotes the continued purchase by a consumer of particular goods or services, using a simple promotion. As an example of this concept, some brands of dog food contain a paper coupon that is found near the bottom of the bag when it is almost empty. The consumer, doling out the last scoops of dog food, discovers the coupon at the opportune moment when buying another bag is necessary. The coupon is an economic incentive to choose the same brand of dog food over that of a competitor, and the message is delivered each time the consumer is about to make his or her next purchase. The methodology of causing a consumer to create an economic addiction to a medicine is more complex: potential consumers are presented with a medical condition, usually not life threatening, which they may have. Then they are shown a medicine that may relieve, but not cure, that condition. Then they are told, "Ask your doctor if [NAME OF DRUG] is right for you." It is important to keep in mind that these are not patients who went to their doctors, seeking treatment, but rather were 'advertised to' and responded by making an appointment to consult a physician. It is then left up to the physician to 'close the sale' and to that end, physicians are assailed by pharmaceutical company representatives who tout various drugs, even providing doctors with free samples to distribute to their patients. In keeping with the 'dope pusher in the schoolyard' model of the previous century, 'the first one is free.' This marketing model quite baldly attempts to make someone decide that they need the product, which effectively makes them dependent on it—*i.e.*, addicted to it—economically, if not physically.

Key factors identify venditiogenic addiction type drugs. First and foremost, they are not medicines that cure, but rather, medicines that 'treat.' In other words, they relieve symptoms, but do not improve the patient's health so that the problem being medicated goes away. Second, they are carefully priced: high enough to be immensely profitable, but not so high as to cause consumers to reject them. Thus, for one or two dollars a day, a patient can be free of an annoying or embarrassing symptom, such as the runny nose and itching eyes due to allergies. The consumer decides that to be rid of the symptom is worth the expenditure of one or two dollars per day.

The third characteristic is not a pleasant one for pharmaceutical companies, but is somehow inexplicably easily tolerated by consumers: side effects. Almost invariably, these 'maintenance' drugs have lists of annoying side effects, such as loose stools, flatulence, nausea, diarrhea, blurry vision; and occasionally, in a small percentage of cases, there are grave side effects, side effects that can cause incredible debilitation or even death. And yet the medical marketplace—comprised of doctors and consumers alike—considers these risks to be acceptable. One has to wonder, what would Hippocrates have thought about all this? Does this not violate the spirit of *Physician, do no harm?*

The physician's answer to that would be, *"All medicines pose some risk."* This concept, that any drug therapy has the potential to cause harm, now seems to be used to justify the prescribing of any drug for any condition. The idea of risk, however, is not a vague and incalculable notion. Financial analysts and traders, whose business is based upon gambling gain against loss, have a tool they use to make reasoned decisions about investments, the *risk:reward ratio.* They quantify financial exposure and measure it against potential profit to determine whether a given venture is financially too dangerous. The same principle may be applied to drugs in a *risk:benefit ratio.* If the *benefit* of a drug therapy is less than its potential *risk,* then the therapy should not be employed. The idea seems almost childishly simplistic, and yet many physicians appear not to use this type of criteria-based decision-making when prescribing drugs. My wife suffered from Temporomandibular Joint (TMJ) pain and was given a Non Steroidal Anti Inflammatory Drug (NSAID) to alleviate the condition. While the drug relieved the pain and clicking in her jaw, it resulted in her being hospitalized with acute pancreatitis, an attack so severe it was not thought that she would survive. In order to alleviate a non-life-threatening symptom, she was given a medication with the known potential of causing a life-threatening health problem. Even a casual analysis of treating her original condition with NSAIDs would show an unfavorable *risk:benefit* ratio.

Prescribing the off-label use of antipsychotic drugs such as *Seroquel, Zyprexa* and other similar compounds, to treat insomnia appears to be a growing trend among physicians. A Knight Ridder analysis of prescribing practices during the year ending July 2003 reports that a full 90% of prescriptions for the anti-seizure drug *Neurontin* were for off-label use, meaning that it was used to treat the conditions for which it was approved only 10% of the time. Not only are such drugs quite difficult to discontinue, they carry the risk of causing not only diabetes, but a condition known as *tardive dyskinesia,* characterized by uncontrollable, spastic movements of the muscles of the face and body. Tardive dyskinesia can be permanent, persisting after the discontinuation of the medicine that caused it. To gamble *permanent* damage to a patient in order to solve a *temporary* problem such as insomnia seems like rather poor judgment in weighing risks and benefits. We are urged by countless advertisements to ask our doctor if a particular medication is right for us. It seems hardly certain that our doctor is, in fact, the right person to ask. Even if a physician deems a drug therapy to be appropriate, he or she should provide us with *all* of the information about risks associated with it. The concept of 'informed consent' relies entirely upon the degree to which we are informed. Had my wife known she would be risking her life, she may not have chosen to take an NSAID for her jaw pain. If I had been told that by taking benzodiazepine I would risk losing years of my life to a debilitating dependency on the drug, I would certainly have refused it. Now that I understand better the economic realities of drugs, of pharmaceutical companies, of governmental regulatory agencies, and of doctors' prescribing habits, I will never again place my trust blindly in others. It's my body. I'm responsible for it. Should a physician suggest a medical treatment to me in the future, I will do exhaustive research into the actual safety and efficacy of the treatment before I give my consent.

Chapter Sixteen
Haywire

As the months passed with glacial slowness while I tapered my use of Valium, the amount of benzodiazepine in my bloodstream and tissues gradually diminished. From 60 mg down to 40 mg, I cut my dose by 5 mg each week. From 40 mg down to 20 mg, I reduced the amount of the cut to 1 mg each week, and from 20 down to 10, the cut in dosage was 0.5 mg. The anxiolytic action of the benzodiazepine was diminishing with the dose; I didn't feel that the drug was having much of an effect by itself now that I was taking so much less of it. I took my dose at night, before bed, but that was mainly out of habit. While my full dose of Xanax—back when it still worked—would essentially knock me out and send me, unconscious, into sleep, the reduced amount of Valium certainly did not have that result. I viewed it solely as the means for getting off of benzodiazepine safely, in gradual increments, so as not to damage my nervous system in a way that would cause protracted problems.

Meeting each day was a challenge. Many of my normal systems felt as though they had gone 'haywire.' My visual perceptions were often distorted. My sense of balance was askew, causing me to feel always as if I were leaning forward, or listing to one side or the other. Occasionally, I would attempt a simple exercise to help the condition: I would raise one foot a few inches off the floor and try to balance on one foot. This would invariably invoke a sudden flush of vertigo and I would quickly put my foot to keep from falling over. A lifelong reader, I found that I could no longer read books. During the first part of the discontinuation process, my cognitive impairment was so great that I simply could not make sense out of anything I read. I could not connect the ideas to each other—the effect was that each sentence appeared to exist solely by itself, unrelated to sentences that preceded or followed it. After a considerable amount of time had passed I found I was once again able to comprehend the text in books, but a new problem arose. Regardless of the time of day, if I began to read, I would drop off into sleep after only a few pages. The sleep was very short, of only a few minutes' duration, but as a result, falling asleep for the night later became an impossibility. I tried to use this to my advantage by attempting to read when it actually was an appropriate bedtime, and thus foiling the insomnia discontinuation had brought on. But, once again, I would nod off for only a few minutes, then wake up, unable to sleep again until the following night.

The intensity of fatigue waxed and waned: at times I was capable of doing the rudimentary tasks life requires, but at others even the simplest things were beyond me. The effect of being so disabled was that I not only felt marginalized, I was, in fact, truly marginalized. I came to recognize the ironic verity of the term, *invalid:* as my reduced capability as a human being left me feeling literally 'invalid.' I could not contribute to life the way a person with even nominal health would do. To avoid stress, I avoided the society of other people. It seems unimaginable, but to someone who is debilitated by what is essentially a nervous disorder, the amount of energy it takes to interact socially, even with a good friend, can seem immense. Like many others who have problems with benzodiazepine, I became functionally agoraphobic—although I don't know whether I actually had developed the psychological condition of agoraphobia or whether I simply had all the symptoms of the condition. The point is merely academic, though, as the result was the same: I was housebound. I minimized my contact with others. When being with other people was unavoidable, I 'faked it,' feigning being okay. To avoid unnecessary explanations about benzodiazepines, I told people who asked simply that I had Chronic Fatigue Syndrome and left it at that; unless, however, I thought someone might be at risk of going through benzodiazepine withdrawal themselves, in which case I would share what had happened to me and what I was doing about it.

As ever, the most valuable camaraderie I found was at the bulletin board at *www.benzo.org.uk*. The administrators of the website were unflagging in their support for completing the daunting task of getting off of the drugs, and the companionship of others who either were tapering with Valium or who had completed the process was deeply comforting. The ordeal was so difficult, and the withdrawal phenomena so disturbing and peculiar, only someone who had experienced it himself or herself could relate to what I was going through.

One of the more distressing effects of benzodiazepine withdrawal was the change it wrought on my libido. This complaint appeared to be an almost universal problem among people with benzodiazepine withdrawal issues. During the intense period when I was in interdose withdrawal while still taking Xanax, I was so minimally functional I wasn't even capable of noticing that my libido had disappeared. After a month or more of tapering with Valium my general condition improved to the point that my sex drive returned—but it, too, was one of the systems that had gone 'haywire.' Only some of the natural manifestations of libido had come back after I had stabilized on Valium, and I found myself to have *anorgasmia,* the inability to achieve orgasm, a condition which lasted throughout the entire first year of the taper. I was still mentally incapacitated, which only added to the confusion that changes to my sex drive caused.

It is only reasonable that GABA down-regulation would result in sexual dysfunction. Like the urge to eat or sleep, sexuality is a core primal function and therefore very much a part of our most primitive nervous system's normal operations—which become disrupted by changes in the regulation of neurotransmitters such as GABA. Sexuality is so intensely personal, however, that our reaction to anorgasmia would naturally be more complex than to insomnia, for example, though they both may result from benzodiazepine withdrawal. The ramifications of sexuality extend far beyond either procreation or pleasure. Since who we believe we are as sexual beings is fundamental to our sense of identity, a major disruption of that belief would necessarily undermine our concept of self.

While we perceive its effects at the personal level, the actual cause of down-regulation of libido, however, may well reside in our own biology, deep beneath the layers of highly sophisticated consciousness that comprise our individuated personalities. We may be human, but we are yet animals, and must operate in accordance with the innate principles that govern animal life, the most compelling of which are related to survival. In the natural realm, life is a rather uncertain proposition, dangerously lacking in security or stability. Finding enough food to eat is a perilous endeavor, fraught with danger, and requires that an animal perform at the peak of its abilities.

Wolves in the wild, for example, when fighting over issues of turf or dominance within the wolf pack hierarchy, snarl and snap fiercely at one another—and yet hardly ever land an actual blow. Most of their conflict turns out to be theatrical posturing. Their behavior reflects the underlying reality that survival is so tenuous, a single wound, even a minor one, might diminish the capacity of the animal to meet the difficult challenges of providing food for itself and the pack. So narrow is the margin between success and failure in their harsh environment, a lapse in securing nutrition for just a short period of time might easily cause an animal to enter a downward spiral from which it is impossible to recover. Since the wolf pack functions as a group, the loss of one of its members as a provider affects them all. Wolves appear to understand this instinctively and conduct their internecine conflicts in such a manner as to result only very rarely in blood being drawn. Although all of the adults in a wolf pack are biologically capable of reproduction, only the fittest of both genders, the Alpha Male and Alpha Female, copulate and breed. When their offspring are born, all members of the pack join in providing food as well as nurture for the wolf cubs. This behavior suggests that it is understood that available food supplies are too scant—and the natural risks too great—to support a multiplicity of offspring by a multiplicity of breeding females, so the limited resources of the entire pack are combined to insure the survival of the progeny of just one pair of animals.

When an animal is operating efficiently, effective in providing food and security for itself and its offspring, it can be regarded as being in a *productive mode*. Should the animal become injured or sick, however, it would enter a *conservative mode,* finding a place of relative safety in which to rest and recuperate. To attempt to hunt when its prowess is lessened would not only expose the animal to environmental risks, the higher level of activity would deplete energy from the healing processes, increasing the possibility that death might ensue. While in a conservative mode, an animal's urge to hunt is supplanted by a *disinclination* to do so. The disinclination to engage in sexual activity may be even stronger because such behavior would put more than just one animal at risk: if an injured, ill, or otherwise compromised male were to impregnate a female, he would have to be even more productive than usual to provide for the mate and the offspring. A female in an unhealthy condition would be less able to meet the taxing demands of nurturing her young. Since lust and libido compel animals toward behavior that would tend to result in pregnancy, the limbic brain, wisely, decreases these functions drastically.

The programmed productive and conservative modes occur at a very deep level, where we are not complex persons but, rather, simple biological creatures. Changes in mode, however, are reflected in all levels of our awareness. While the instances in our lives where we are actively seeking to produce children are quite rare, it is evident that maintaining the option to do so is a driving force in the human experience, and informs much of our behavior, from personal grooming and choosing what clothes to wear to how we conduct ourselves in the presence of others. Although the condition of having down-regulated GABA functions as a result of benzodiazepine withdrawal is neither a disease nor an injury, the degree of debility it creates can invoke the most pervasive of conservative states over a protracted period of time. While we are so thoroughly impaired, to conceive offspring would be a profoundly inapposite thing to do. Perhaps this creates a biological injunction against orgasm, and against libidinous impulses in general. Even anhedonia, the inability to experience pleasure, can be seen in terms of survival instinct. When an animal's health is compromised, it is vital that it keep its attention focused on the problem until it gets resolved. Pleasure would tend to distract attention *away* from the possibly dire condition; since the consequences of that could result in death, anhedonia would serve the pro-survival function of maintaining focus.

During the period when my libido was at its lowest ebb my ability to think clearly was so compromised that I could never have sorted out that there were possible biological underpinnings to the condition. I was experiencing what people in benzodiazepine discontinuation refer to as 'cog fog,' a perplexing state of dulled cognition in which thoughts tend to be muddy, disconnected, easy to lose track of. It was only much later that I was able to realize that my reaction to my diminished libido was, itself, somewhat disturbing at times. Confused, perhaps, about feeling a loss of my identity through libido, I was unable to discern what sexual ideation was appropriate and what was inappropriate, neither according to external societal standards nor, more importantly, to my own. I felt embarrassed and ashamed, the sense of having lost my bearings only adding to the emotional distress I was already feeling. My disorientation persisted for what seemed like an endless amount of time.

Chapter Seventeen
Thinking and Feeling

The psychological effect of 'cog fog' was devastating. Whatever measure of intelligence we are given, we tend to feel that that is our due, something that belongs to us. Therefore, to have it eroded to a marked degree is deeply disheartening. I was grateful that I was not delusional, and that I was able to think rationally, to consider pieces of information and make determinations about them. But the manner in which my mind functioned had obviously been profoundly affected by benzo withdrawal. In a word, I felt *stupid*.

In addition, there was often a disturbing sense of distance between my thoughts and reality, as though 'I' were in one room and reality in another—with an additional room in between them. Fortunately, the disconnection from reality I was experiencing was minor compared to what I knew others were suffering as a result of benzodiazepine withdrawal. Its more acute form, known as *derealization,* can cause people to feel almost entirely dissociated from existence, detached from their experience of themselves, as though reality were no more than a movie of itself. The world may appear to be distant, dreamlike, or even as though objects in it were distorted and flat, or made out of clear, translucent jelly. An adjunct component to *derealization* is *depersonalization,* in which people lose their sense of identity, often perceiving themselves as two-dimensional 'cardboard cut-outs' acting out their lives without participating in or connecting with them. While not as intensely as others did, I certainly experienced elements of depersonalization: at some times I would see myself in the wreckage of my former life and feel despair at all I had lost and at what had befallen me, but at other times I would seem a mere observer, watching myself living out a wretched parody of life without feeling any emotional reaction to my diminished state. I was like a ghost, haunting my own house. I performed the same ritualized behaviors day in, day out, without spontaneity, without any sense of 'flow.' That I could function in such an appalling manner and *not* be appalled was indicative of the degree of my depersonalization.

When I saw that other people at the online bulletin board at *benzo.org.uk* played simple word and number games the wisdom of doing so became apparent. Anything that would exercise cerebral processes should tend to strengthen the neurochemical underpinnings of those processes. My ability to form ideas had been drastically diminished, and my command of words—which reflect ideas—had been severely compromised. I decided to focus on words as a means to restoring my damaged intellect. My first attempts to play word games, however, were discouraging, so I focused on a memory game of the 'concentration' type I found at a website called *toadgames.com.*

Many pairs of symbols are concealed behind squares in a grid. Clicking on one of the squares would momentarily reveal the symbol that lay behind it, and if both squares concealing a particular symbol were clicked, the symbol gets eliminated. As a player has to recall where the symbols have been seen, the game promotes memory.

FIG. 5 : *A 'concentration' game*

toadgames.com has games of many types, each of which can be seen to address a different type of mental function, from memory to reasoning to hand-eye coordination, and I employed many of them in my rehabilitation. The fatigue component of benzo withdrawal meant that I rarely had much physical energy to do anything more ambitious than merely sitting in a chair, so the ability to turn those tedious hours of sitting into something that might actually do me some good was deeply appreciated. I would play solitaire, getting little more from it than the satisfaction of being able to follow a simple procedure, but, considering my low state, even that was helpful.

Gradually, my repertoire of games increased to include ones that required strategic thinking and greater focus. Next, I began to play word games, most particularly *Word Noodle*, which required being able to form words out of adjacent letters randomly placed in a grid.

JACK HOBSON-DUPONT 135

FIG. 6 :   *'Word Noodle' at toadgames.com*

I found other word games to be helpful as well, such as *Word Find*, where the player is presented a list of words which occur in a vast field of letters and must find them, and *Word Jumble*, a syndicated game that appears in newspapers but can be played online.

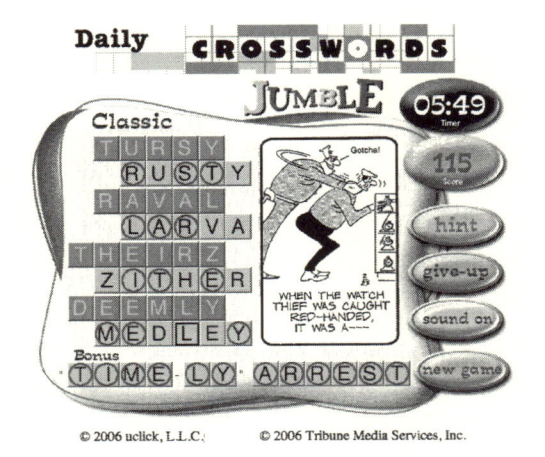

FIG. 7 :   *Internet version of 'Word Jumble'*

With my thinking impaired, performing well at word games proved to be difficult. I struggled with such games and improvement in my scores occurred only gradually. The opportunity to focus my mind by engaging in cognitively stimulating activity, however, was a valuable tool in my recovery. Higher scores in word games—when they did occur—provided an objective way of measuring my progress with other intellectual abilities. While word games were helpful to me, *toadgames.com* abounds with many other types, games that use strategy, require visual or graphical skill, or even hand-eye coordination. Any such pursuits would tend to enhance cognitive health.

Not only can the intellect become dulled during the ordeal of benzodiazepine discontinuation and withdrawal, emotions may be dulled, as well. Many people in benzo withdrawal report experiencing *emotional blunting,* where their normal emotional reactions have been obtunded, often to an alarming degree.

Professor Ashton, in *Benzodiazepines: How They Work & How to Withdraw,* says,

> *'Emotional anesthesia', the inability to feel pleasure or pain, is a common complaint of long-term benzodiazepine users. Such emotional blunting is probably related to the inhibitory effect of benzodiazepines on activity in emotional centres in the brain.*
>
> *Former long-term benzodiazepine users often bitterly regret their lack of emotional responses to family members—children and spouses or partners—during the period when they were taking the drugs.*

Emotional blunting is a widely recognized symptom not only of the use of benzodiazepines but of SSRI antidepressants, as well. A study reported in the June 2002 issue of *The International Journal of*

Neuropsychopharmacology revealed that,
> *80% of patients with SSRI-induced sexual dysfunction also describe clinically significant blunting of several emotions. Emotional blunting may be an under-appreciated side-effect of SSRIs. . . .*

Given that disruptions of sexual expression are quite prevalent in SSRI use, the number of patients with emotional blunting is, therefore, significant.

The symptom can extend throughout the discontinuation period and beyond, much to the consternation of those who have it. At one point during my taper, a woman wrote me an e-mail when a family member died. Her grandmother had raised her and been her primary source of love and unconditional acceptance throughout her life. When the grandmother died, she 'felt nothing,' and had to 'fake it' during the funeral ceremony. A year passed before she could feel any emotion whatsoever about the loss of so beloved a person in her life. It is ironic, perhaps, that the grief of the death of a family member would be met with emotional blunting, since many doctors prescribe benzodiazepine to people who are coping with death. The doctors, with understandable compassion, seem to seek only to relieve the great distress of the survivors, but the practice transforms the deep sadness of a tragedy into a medical problem. And worse, many people have reported that being 'treated for grief' was their introduction to benzodiazepine. In due time the impact of the loss naturally subsided, but the prescription had persisted—they then found themselves habituated to being on a tranquilizer for the long-term, with all of the difficulties of rebound, tolerance, dependence, and the difficulty of discontinuation that entails. To be medicated in order to avoid experiencing deep feelings quite logically leads to the emotional disconnection from reality that so often occurs as a result of both the use of—and withdrawal from—tranquilizing drugs. It would seem far healthier if we, in our cultural sophistication, had not used this method to evolve past wailing and sobbing at the passing away of loved ones, treating grief as though it were a disorder rather than a component of normal life.

Chapter Eighteen
Therapy During Recovery

Emotional blunting is the type of problem for which it would usually be wise to seek therapy. Many of the sequellæ of benzodiazepine discontinuation would also be found in that category, and yet, given the nature of what recovery from benzodiazepine-induced damage actually is, talk-therapy for people with benzodiazepine difficulties may not a viable component of recovery.

Benzodiazepine-caused problems have nothing to do with psychological conditions, and bear only a tangential relationship to them. As Professor Ashton's research clearly points out, the sole problem to be overcome is for the central nervous system to reestablish the affinity of its neural receptor sites for the naturally occurring brain chemical, GABA. This is not a psychological issue. It isn't even a medical issue, in the sense that it is not a 'disease'. Rather, it is more akin to a mechanical malfunction than anything else, and apparently the only remedial elements which materially affect recovery are the passage of time combined with a lack of any further trauma interfering with the brain's slow recuperation and restoration of its natural biochemical affinity for GABA. Therefore, costly psychological counseling is frankly inappropriate in such a process as it could hardly produce effective results. Since the source of the disturbing phenomena is physical, it would be irrational to attempt to correct them by addressing the psyche. Also, the psychotherapeutic method relies upon a patient being able to respond to feelings. The therapist guides the client through emotional explorations, often of painful areas of experience. The only way the client can tell that an issue has been resolved is if how they feel about it changes. But in the case of someone who has been exposed to benzodiazepines, how they feel may well be influenced profoundly by the lack of efficacy of the GABA in their bodies, so even though they may resolve a problem psychologically, they might never know it because they could be physically incapable of having feelings of relief such a change would normally bring about. As such, to process emotional information while under the influence of benzodiazepine withdrawal phenomena might so undermine the therapeutic dynamic as to be dangerous to emotional health.

One of the people I met and become close with at *benzo.org.uk* was seeing a psychiatrist when a too-rapid withdrawal from Xanax severely affected her mind. One of the things he recommended as a way of recovering from her state of depersonalization and derealization was to get her boyfriend to read to her from Danté's *Inferno* each night. I suggested she tell her therapist that if he thought reading from Danté's *Inferno* could bring about the massive damage to her mental abilities she was experiencing, then reading from it might also be powerful enough to reverse the condition. But since it wasn't, the idea that it could effect a cure was ridiculous.

A published author and doctoral candidate before withdrawal from Xanax affected her cognitive abilities, the young woman found herself having intrusive thoughts of killing herself. She wrote a suicide note and brought it with her to her psychiatrist. Standard procedure for a therapist when a client brings up suicide is to admit them to a hospital (in large part to avoid being held responsible for negligence should the person actually commit suicide.) Therefore, the psychiatrist walked the woman to a hospital to check her in. But along the way, he had her stop at an ATM machine to extract the money to pay for her last visit—so that if she had gone through with her suicidal ideation and killed herself, the psychiatrist would not have been 'out of pocket' for the expense of her final session. It was not until months later that the woman regained enough cognitive ability to realize that the doctor's treatment of her had been somewhat monstrous.

Instead of the role of 'therapist' a case may be made for the role of a 'benzodiazepine counselor,' *i.e.,* someone to help people deal with the bewildering assortment of problems that recovery from benzo withdrawal can entail. While people may find comfort from such counseling, as well as good advice based upon the experiences of the counselor with others having undergone the same process, the process would not constitute actual 'therapy' in that the counselor cannot actually help the brain recover its affinity for GABA.

There are many techniques that victims of exposure to benzodiazepine can do in the area of promoting their own psychological health which will help them cope with the process of recovery: developing a positive mental attitude, learning positive self-talk techniques, learning to differentiate between 'false reasoning' and reasoned reasoning to help remain grounded in reality as much as possible during an often bewildering process. Why these techniques are important, though, is not because they will have any effect upon the efficacy of GABA, and, more to the point, has little to do with psychology itself: they help because they tend to reduce the amount of excitatory stress, and lowered stress would promote an environment in which the brain can better rest and recover.

Looking at my life before benzos, I see that it was characterized by feeling optimistic and cheerful—in spite of the fact that I was often fairly unfulfilled in my circumstances. I read something once which struck me as absurdly simple yet profoundly true: "To live a happy life you must think happy thoughts." Considering that idea, I realize that it's at the very root of what a 'happy life' actually is, simply one in which someone thinks predominantly happy thoughts, regardless of external circumstances, positive or negative. Unfortunately, in addition to the situational depression that undergoing benzo withdrawal often imposes just by the very nature of its disruption of 'normal' life, the down-regulation of GABA also can cause purely neurological dysphoria and anhedonia, rendering it neurochemically impossible to think happy thoughts, even for those prone to having them. Happy thoughts may well be the products of a natural *feeling* of happiness; they may simply never be formed at all if the ability to have a feeling of happiness is compromised neurochemically.

Chapter Nineteen
Compassion Burnout

In the absence of a trained professional to give counsel during benzo discontinuation, most people turn to their spouses, family members and friends for support through what may well be the most difficult ordeal they have ever faced. That is unfortunate because the likelihood of such support being sufficient is rather small. The problem is not that such people are lacking in kindness or sympathy, but that the emotional needs of those who are enduring the effects of benzo withdrawal can be so vast as to consume whatever care is offered, and that the effects themselves can be so numerous, so bizarre and horrendous, that no one who has not experienced them could understand them well enough to empathize. Perhaps even more importantly, the sheer amount of time it takes for withdrawal and recovery can extend so long as to cause what is known as *compassion fatigue* or *compassion burnout* in caregivers.

We are accustomed to having loved ones become sick and require our attention to help them recover. It is a familiar part of civilized life, with unconscious expectations by both the caregiver and the receiver of care. There are usually two paths this can take: either the sick person responds to nurture and recovers, or they die, and the caregivers grieve their loss and then get on with their lives. In either eventuality, there is a *time scale* for the giving of care. Chronic illnesses requiring continual care by family or friends—as opposed to professionals—is exceptional. While some people may well discontinue benzodiazepines without much difficulty, for those who do have difficulty, the process is often almost absurdly lengthy, taking years to accomplish. During that time, month after month elapses in which there is no improvement, no sign that there is even any hope of improvement. Caregivers come to despair of having any favorable response, leading them to feel that their efforts are wasted, their attempts to help futile. They may come to resent strongly the fact that the benzo victim appears never to respond to their best efforts at providing support. Naturally, they have no way of knowing that benzo recovery often takes an inordinate amount of time to occur, so they may perceive a person in withdrawal as being a malingerer. Family members and friends may tend to project their own ideas of healthy living upon a benzodiazepine user, urging that fresh air, exercise and keeping busy would effect a positive outcome. Or, they may impose the only template they understand—the typical 'addiction' model—onto benzo withdrawal, believing that the problem represents a lack of will power, self-restraint and moral fiber. Since benzodiazepine discontinuation is so poorly understood by doctors and other medical professionals, it is not surprising that it is misunderstood by well-meaning but misguided loved ones, too. Without intellectual grounding in the fact that the sole difficulty with benzodiazepine discontinuation lies in the down-regulation of the GABA-α receptors, caregivers would have no way of understanding that to have a positive attitude and a healthy lifestyle could ever only secondarily affect the condition itself or the speed of recovery from it.

In my own case, I was in a state of debilitation for about six months from having reached tolerance on Xanax. Once I learned that Xanax was the cause of my condition, I crossed over to Valium and began tapering the drug, a process that took an additional two years. After finally getting off Valium altogether, my condition improved only marginally over the ensuing months. Therefore, I spent three years in a state of debility and during that period, virtually every time my wife asked me how I was feeling, she received an answer that could only have been deeply discouraging to her. That amounts to over one thousand days of hearing bad reports. While she did not suffer the actual horrors of what I was going through, they had a devastating effect upon the quality of her life, too. Whatever vision she had of our time together was destroyed—I was incapable of going to parties or on outings with her, taking vacations together, or even of doing much more than watching television with her in the evenings. She no longer had an active, dynamic husband; I was reduced to someone sitting on the couch, day after day after day after endless day.

After awhile, for her own survival, my wife learned to disconnect herself emotionally from my seemingly perpetual state of gloom and despair. She would ask how I was feeling, but when I would reply, she would contain my response in such a way that it would not drag her down into a gloom and despair of her own. It was essential that she do this—after all, she had to be the sole provider in our family, as well as being the source of all of the assistance our son needed during his senior year of high school and his first years of engineering college. I simply wasn't capable of offering very much at all. My wife had reached compassion burnout, and had responded in the best way she knew how. Had she not, our whole family would have been brought down by what had happened to me.

I spared my son from as much of my ordeal as I could, mainly by keeping myself sequestered away as much as possible. I didn't seek his support, since to have engulfed him in my peculiar health problems would only have confused him, and more importantly, would have reversed the dynamics of the parent/child relationship structure and burdened him inappropriately. He knew that I was quite unwell, that it was because of an addictive drug, but that my condition was not life threatening. I did not wish to intoxicate him with much information beyond that. I treated my friends in a similar way, letting them know what had happened to me, but not involving them in what I was going through. In truth, their care would have been of little actual help to me. There was nothing anyone could do that would actually have caused me to feel better, since neither their efforts nor my own could actually restore GABA to its full efficacy.

While my wife provided material support for me to endure the lengthy process of getting off of benzos, the prime sources of emotional support for me were people who knew only too well the sort of thing I was going through: other victims of benzodiazepine withdrawal. I found them through Ray Nimmo's website, *benzo.org.uk*. Sharing a common misfortune, we got to know one another well through posting messages in *benzo.org.uk's* online bulletin board. Withdrawal can present so many unusual symptoms that, without others to corroborate our experiences, we might never know they were all related to benzodiazepine use. For example, some of the 'old timers' who had discontinued the drug in the information-starved era before the Internet had had their teeth removed because of disturbing perceptions that their teeth were vibrating or felt metallic or 'rubbery.' Only when the sensations continued after the teeth were extracted did they conclude that the symptoms were related, not to dental problems, but to benzodiazepine withdrawal. Once there was a way to share such information among individuals efficiently, *i.e.,* via websites and bulletin boards, people experiencing strange dental sensations could easily find good advice against unnecessary tooth extraction as a remedy.

Since people in benzo withdrawal are often anxious, isolated, housebound and agoraphobic, the idea of attending a real-world meeting of fellow sufferers would hardly seem practicable. One major virtue of an online 'meeting' is that someone can participate from the relative security of their own home, merely by logging into a website. Another virtue is that an online community has no geographical constraints—based in the United Kingdom, Ray Nimmo's bulletin board had members located all over the world. Individuals are therefore exposed to a larger pool of experiential information than they would at a local meeting. Not only do the members of such communities provide continuous support, the administrators are often people who have dedicated their lives to helping others in the struggle to get safely off of benzodiazepine drugs, without pay or recognition for their efforts.

Discontinuation from benzodiazepine is often a lonely undertaking. The ability to connect at any time during the day or night to a forum where like-minded people may be found can be profoundly comforting. Where family members and friends might understandably fail to empathize with the unique problems of people in benzodiazepine withdrawal, dedicated online communities are able to provide support, compassion and understanding over the excessively long span of time that withdrawal sometimes requires.

**Chapter Twenty
The Lower Doses**

One of the profound benefits of spending time at an online community of people who were going through—or had already gone through—discontinuation from benzodiazepine was the opportunity to avail myself of the collective wisdom of everyone there. The process was doubtless the most challenging thing any of the participants had ever undertaken in their lives, and by sharing with one another their difficulties as well as their triumphs, a great deal of important information about successfully getting off of the drug was generated.

I noticed that many people reported exceptional problems when they were at lower doses of Valium. While I didn't doubt the validity of what was being asserted, the idea seemed illogical to me. How could there be greater problems when there was less of the drug present in the body? The only reasonable answer was that the problem lay not so much within the primary characteristics of the drug but on the secondary effect it had on GABA regulation.

As my own dose of daily Valium steadily declined, I found myself more and more capable of 'normal' activities, albeit in highly limited form. I had reduced my dosage from 60 mg of Valium down to 9 mg by the time the next summer had arrived, and I endeavored to take advantage of the mild weather. The two preceding winters had been unusually cold, which had added its own form of stress to the internal stress I had been experiencing. I had observed that I tended to hunch my shoulders up against the cold and my muscles had seemed to lock in that position. When I walked, my arms remained stiffly at my sides. During that summer, I went on short walks and kept reminding myself to swing my arms to loosen them up—but without those constant reminders, I saw that my shoulders would tense back up and freeze into their rigid position.

One of my passions in life is sailing, and I am fortunate to live within a few blocks of the ocean. That summer, I launched my small sailboat and kept it in the nearby harbor, going out on it as often as I could. It was a peculiar experience. I was constantly asking myself, "Am I actually enjoying this?" as I sailed in the local waters. Here I was, doing something I knew I loved, but I had no feeling about it whatsoever. I would look at the sunlight glittering on the water, observe the sea birds diving on shoals of baitfish, sense the motion of the boat through the water, all in stark contrast to the many previous months of simply sitting in a chair day and night. And yet, I didn't actually *feel* any pleasure from it. I could only deduce, intellectually, that this was a pleasant way of spending time, that the sights, sounds and smells were a welcome break from being housebound, and that I *must* be enjoying it on some level. But I had no more feeling of enjoyment than if I had been a robot, physically capable of managing the sailboat and 'processing the data' of sensory information that functioning in such a rich environment imparted, yet devoid of any emotional reaction to it. The only exception was an occasion when I took a young friend out on the boat to teach him the rudiments of sailing. As I often did when I was in the company of others, I was 'faking it', pretending to be normal so as not to impose my own somewhat bleak and bizarre reality onto other people. We had to navigate through some tricky currents to get to a neighboring island, and there was a certain exhilaration in successfully reaching the distant shore. I was so engrossed in the procedures and the company of my friend, I actually lost myself in the experience and realized, when it was over, that I had actually had a good time, had actually enjoyed myself. The next time I went out on the boat alone, thinking that I might once again have a pleasurable sail, I found that my emotional blunting had returned. I persisted in going sailing throughout the summer, however, since I believed that while I didn't really enjoy it, sailing a boat was preferable to sitting at home and would, at least, support the neurochemistry of 'normal' behaviors.

The phenomenon of having a day in which I felt fairly okay—in the midst of an endless parade of days where I felt absolutely terrible—is peculiar to benzodiazepine withdrawal, and quite possibly unique to it. During the post-benzo period of recovery, many people report having 'windows' where, usually quite suddenly, they will simply feel good. The window may be brief, just lasting a few minutes or an hour or so, and will inevitably close again, with a return of the previous malaise. But the fact that it has happened is deeply significant: it is proof positive that recovery can take place, that someone truly can feel good again. Those who have experienced them usually say that recovery consists of a series of these windows, where the time that the window is open increases in length and the interval between their occurrences becomes shorter and shorter until, finally, the window simply remains for an indefinite period. Viewed neurochemically, the phenomenon would appear to be an instance where the preponderance of the GABA receptors' alpha subunits is operating normally, their affinity for attracting GABA restored to its normal level of efficacy. The window closes because the body evidently cannot sustain that state for long, but, over time, continues to regain it again and again until the functionality has been reinstated.

To someone who has grown inured to the misery of benzodiazepine withdrawal over months or, perhaps, years, the sudden appearance of complete respite from the affliction must appear to be miraculous. Since the nature and severity of symptoms that can affect people in benzo discontinuation are perceived subjectively as 'illness,' to experience a window would have the same impact as if a serious illness had gone into sudden remission. One cannot even imagine the impact on a blind person of unexpectedly having an hour of sight. And although it would be deeply disheartening for that window to close again, signaling a return of the previous condition of discomfort, the significance of the window would remain: since it happened at all would mean that it can happen again, that recovery is actually possible.

That idea, that recovery is possible, is often hard to believe when someone is in the throes of withdrawal with its catalogue of agonies and indignities. For that reason it is especially helpful when former members of an online benzo community stop back for a visit and to tell people still tapering about how well they are doing after having made it past the residual effects of the drug. Such encouragement does much to allay fears that the process will never end; considering the time frame of getting off of benzos—years, in my case—it certainly seemed at times that it was endless and that my efforts were hopeless.

As that summer wound down, my dose diminished accordingly. I found, however, that as the dose got lower, I really began to 'feel the cuts.' Almost everyone at the *benzo.org.uk* bulletin board reported that they reacted to each cut, usually in a fairly predictable manner. Because Valium has such a long half-life in the body, it would usually take two or three days for anyone to feel the effect of a cut in dose, but then they would typically have a number of rough days before they would stabilize at the lowered dosage. After a few days at that level it would be time to cut again, and the cycle would start over. I had certainly had withdrawal symptoms throughout the time it had taken me to get from 60 mg of Valium down to single digit doses, but they never appeared to follow any pattern. Now, however, they began to. I planned on switching from making once-per-week reductions of 0.5 mg to every two weeks, when I reached a total daily dose of 5 mg. That would give me more time to stabilize between cuts.

Some people at the *benzo.org.uk* bulletin board felt that my tapering schedule was far too slow. After all, it would take me seemingly forever to get off the drug. I had already been tapering for over a year, and it would take another year and a half to complete. It seemed bizarre to me that when I began the program, I could reduce the dosage by 5 mg per week without really feeling much difference. Had I been able to reduce by that much all along, the whole process would have taken only twelve weeks. Professor Ashton had developed her methodology for getting people off benzos by running a clinic in Britain, and had determined that it was best not to reduce the dosage by more than a certain percentage of the overall amount. Therefore, when I was taking 60 mg of Valium per day, a 5 mg cut represented one twelfth of the total dosage, or 8.3%. A cut of 5 mg, however, when I was taking 10 mg would represent half of the total dose, 50%.

The body would have reacted vehemently to a sudden reduction by 50% of the benzodiazepine it was accustomed to.

The slower schedule I had developed was intended to be gentler on my nervous system. Having already spent considerable time in a state that felt to me like being in Hell, I was determined to avoid as much stress to my nervous system as possible. I had read the accounts of others who, for one reason or another, had gotten off of benzodiazepine quickly; only rarely did an accelerated pace turn out well for them. Professor Ashton had pointed out that the tapering schedules she had devised were not hard and fast rules but, rather, recommendations, and that each person might tailor the schedules to suit their particular needs. The only area in which I intended to stray from her counsel was in the very last phase. She recommended 'taking the plunge' by discontinuing directly from a 0.5 mg daily dose. From my long observations of the difficulties of others, I decided that I would taper down further even from that miniscule amount of the drug, again, for the purpose of avoiding stress to my nervous system.

As I reached the 5 mg dose, however, rather than slowing down at that point I continued at the same pace, even though I had begun having stronger reactions to the cuts. I simply didn't want to lose the momentum I felt I had, and I was discouraged by the thought that, even at a small dosage of 5 mg, I still had so many, many months to go before I would be free of taking the drug. Had I thought I could physically handle simply quitting, I would have done so. If only discontinuation of benzodiazepine had been a matter of will power! I found my dependence upon it so loathsome, I would gladly have simply stopped taking it and toughed it out. But, fortunately, I had access to plenty of information about people who had done just that, and the results were usually disastrous. In some cases, those who had discontinued too abruptly would try to reinstate their usage of the drug, but that, too, never seemed to work, even if they increased the dosage. The problem, as usual, was not so much the action of the benzodiazepine itself but its effect upon GABA regulation. Nothing seemed to work except slow, measured, careful tapering—and oceans of time.

Chapter Twenty-One
Fear

As autumn came on, I began to have a new 'w/d,' *i.e,* withdrawal symptom. I had already had neurological problems with the muscles in my legs, but now I began to have an electric *buzzing* in my feet when I woke up in the morning, accompanied by a slight panicky feeling. This quickly escalated, over a matter of days, into waking up to find that the region from my feet up to my thighs and my hands up to my elbows was buzzing, tingling electrically, and numb to the touch. Soon, it spread even farther than that, and if the buzzing sensation made it up into my torso, the panicky feeling would escalate into a state of profound fear. These perceptions would diminish gradually as I went into wakeful consciousness, but they would only diminish, not disappear.

The state of fear was overwhelming, and I couldn't shake the sense of abject terror and horror. It wasn't psychological fear but complete physical fear, and I trembled and sweated, and had trouble swallowing. I found myself in a state of the deepest depression, vastly beyond anything I had ever felt before. Problems with my libido escalated immediately into a state in which I was sex-adverse; in fact, I could hardly bear the company of others in any capacity—and yet, so profound was the depression and fear, I was utterly uncomfortable alone.

I felt completely terrified, and could not even find anywhere within myself where I could feel any comfort at all. And then, the suicidal thoughts began, intrusive thoughts of taking my own life, and I could not seem to escape them. I was never at risk of committing suicide—even while I was having these thoughts, I knew I would not take my own life under any circumstances. I was too well aware of the devastating effect that killing myself would have on the subsequent lives of the people I loved. And yet the suicidal ideation continued, appearing to be generated by the abject state of fear I was in itself.

There was no reason to the suicidal thoughts. They weren't the product of my thinking that my life was unbearably horrible—even though it certainly was. They simply seemed to form themselves, repeatedly, all day and all night. The state of depression I was in was total; there was no part of my consciousness that was spared from its effect, nothing in which I could feel any goodness at all.

Observing myself, I noted that when I woke up, immediately upon feeling the buzzing vibration and fear, I would start to *think*, and the more rationally I could think, the more quickly I could quell the feelings. I saw that any mental state other rationality, the *two-plus-two-equals-four* type of awareness, would bring on the buzzing and increase my sense of panic and despair: daydreaming, idle speculation, random imagination, losing the focus of my thoughts, would all bring it on. I also noticed that the instant I woke up, I would peer intently out my window at the sky, not taking my eyes off of it. I experimented by turning my face away from the window, but found that that increased immediately the intensity of the bad feelings. *Could light have something to do with what was happening to me?*

I have always been prone to Seasonal Affective Disorder (SAD), the phenomenon where the change in light in the autumn causes a depressed mental state, although for me it had never been very pronounced. I did some research on Seasonal Affective Disorder and found that one of its causes is related to melatonin. I had already recognized that when I had taken melatonin as an aid to sleep I had had a bad reaction to it, becoming almost immediately depressed. I learned that melatonin is synthesized out of serotonin—and that neurotransmitter has an unknown but probably influential role in the neurochemistry of depression. As someone with perennial mild depression as a result of benzodiazepine withdrawal, I probably had lower than normal amounts of serotonin available to me already. The production of melatonin was depleting those stores, and the lessened amount of serotonin was quite possibly causing, directly or indirectly, a chemical depression. It was also probably causing the sensation of buzzing and numbness, too. Evidently, one of the functions of melatonin is to desensitize the body during sleep. This must help to maintain sleep by preventing external sensations from arousing the brain from the deeper states of relaxation sleep requires. Normally, we are able to suppress melatonin upon waking up, but in the abnormal condition of my neurochemistry because of GABA down-regulation, that ability was not functioning properly in me. The desperate scramble each morning to dispel dreaminess as quickly as possible by thinking rationally was apparently an attempt to quash melatonin.

I was hardly inclined to use SSRI antidepressants to address the serotonin component of the problem, considering what experimenting with a small reinstatement of Effexor had already done to me. Further research into Seasonal Affective Disorder showed that an effective means of dealing with it was light therapy. I looked into the different types of 'light box' that were available and found one that employed short wavelength blue light which didn't produce any potentially harmful ultraviolet (UV) rays.

I bought the light box and began to use it. At first, I kept it next to my bed and would turn it on the instant I awoke in the morning. It had an array of sixty-six LEDs that would glow with bright intensity. Staring at the blue light produced noticeable effects, beginning with a lessening of the buzzing sensation. It also promoted a slightly jittery feeling, a faint reflection of the 'agitation' I had experienced when taking the SSRI, Effexor, so I assumed this was related to serotonin. Within a few weeks I had stabilized and was no longer waking up to overwhelming irrational fear. The suicidal depression abated, thankfully, restoring me to that state of mild depression that seems common among people discontinuing benzos.

I continued to use the light box throughout the following winter. I moved it down to my computer workstation and would take a 'light bath' each morning for twenty minutes as I began my day. Further research into light therapy had led me to study after study where it was used to great effect in combating depression of every type, not just the seasonal variety. It is widely known that the incidence of depression in the general population is markedly lower during summer, when people are exposed to greater amounts of sunlight daily. What I found curious was that, if light therapy was so efficacious, why didn't more doctors employ it? The low incidence of side effects would suggest that light therapy be the *first* choice a doctor should make when trying to treat depression in a patient. Only if it proved ineffective would a doctor then prescribe antidepressant drugs, with their potential for causing everything from annoying side effects to discontinuation syndromes—and worse.

Light boxes are expensive, typically starting at around $200 in the United States. A doctor could have one on hand, however, and loan it out to patients after instructing them in its use. If it proved helpful, a patient could then order his or her own light box and return the doctor's to make it available for the next patient. The high cost of the light box would be deferred by the fact that, unlike prescription medicine, the device can be used for an indefinite period of time. After the initial outlay of money, there is no further expense. Prescription drugs for depression may cost less initially, but the continued expenditure for them would eventually surpass that of a light box.

Researching this topic, I found an analogue in a type of treatment known as Cranial Electrotherapy Stimulation (CES). First developed in the Soviet Union in the 1950's, CES has long been in use in Europe, where it is usually referred to as *electrosleep*. Electrodes are typically attached to the earlobes and connect to a small, battery-powered device that sends a microcurrent of electricity through them. Since the earlobes are positioned on either side of the head, the microcurrent passes through the brain. This may sound frighteningly like electroshock therapy, but the amount of current used in CES is quite small, typically less than 1.5 milliamperes, about the same as the electrical current naturally occurring in the body.

Cranial Electrotherapy Stimulation has been available in the United States since the 1970's, but its use is surprisingly limited, considering the evidence of its efficacy in treating insomnia, anxiety, depression and pain. A review of over 100 research studies has shown CES to have had beneficial results in 95% of its applications, while presenting minimal side effects—mild headache or localized irritation of the earlobes, neither of which persisted past the treatment sessions. A post-marketing survey conducted on behalf of one CES device, the Alpha-Stim, manufactured by Electromedical Products International, reports:

> *In a survey of 47 physicians reporting on 500 patients, Alpha-Stim treatments produced significant results of at least 25% improvement in 92% of patients for pain management, 94% for anxiety, 90% for depression, 93% for stress, 79% for insomnia, 90% for headaches, and 95% for muscle tension. Nearly half of the patients in all categories had 75 to 100% relief.*
>
> *A survey of 2,500 patients who used Alpha-Stim technology for 3 weeks or more reported similar results. Of 1,949 pain patients, only 7% had less than 25% improvement, 32% had fair improvement of 25–49%, 38% had moderate improvement of 50–74%, and 23% of the patients experience 75–100% improvement. Combining the survey results from the 723 patients reporting anxiety, depression, stress, chronic fatigue, and/or insomnia, only 8% had less than 25% improvement, 24% had fair improvement of 25–49%, 33% had moderate improvement of 50–74%, and over one-third (35%) experience 75–100% improvement.*

and;

> *Only one out of 506 people will experience a mild headache, and one out of 910 will have a skin reaction at the electrode site, usually a minor self-limiting reddening of the skin. There are no other significant side effects reported in over 55 research studies, or in 24 years of clinical and home use. Alpha-Stim is very effective. It provides significant relief for 9 out of 10 people who use it.*

Such a record is in stark contrast with pharmacological drug use with its low efficacy and high rate of often-unacceptable side effects. And yet, CES remains largely underutilized. One reason appears to be the U.S. Food and Drug Administration's efforts to do away with this form of therapy. Cranial Electrotherapy Stimulation equipment is approved by the FDA as medical devices, available by prescription. In the 1990's, however, there was evidently an effort by the FDA to ban them. The agency required that all companies producing them submit their scientific materials for review by the FDA's Neurology Panel, which would make a subsequent recommendation about the classification of the devices. After submission of the requested materials, the FDA assigned the job of evaluating the scientific studies and other documents to a staff member who held a masters degree in mechanical engineering. This employee then asserted that none of over one hundred peer-reviewed studies, masters theses and doctoral dissertations on Cranial Electrotherapy Stimulation was scientifically valid, and concluded that there was nothing to present to the Neurology Panel, effectively forestalling the possibility that the Neurology Panel would certify CES as a viable therapy—as a previous Neurology Panel had done in 1978. The Commissioner of the FDA proceeded with rescinding its classification of CES devices as approved for medical use. One manufacturer, however, sued the U.S. Government over the matter, forcing the FDA to relinquish its attempt to remove CES devices from the market.

While there are many medical devices approved by the FDA, Cranial Electrotherapy Stimulation equipment—and light boxes—pose a threat because their use would, in some cases, obviate the need for certain pharmaceutical products. Once again, it would appear that key administrators of the U.S. Food and Drug Administration have a stake in promoting such products, and suppressing anything that might diminish their sales, even though the mandate of such an agency would suggest that its priority would be toward benefiting the American people, not industry. A government report by the Centers for Medicare and Medicaid Services states that in 2003, health care costs reached US $1.7 trillion dollars, representing 15.3 percent of the nation's Gross Domestic Product (GDP). Those costs then rose by 140 billion dollars in 2004, an increase of 7.9 percent—almost three times the rate of inflation—accounting for fully 16 percent of the total economic output of the United States. It projected that in ten years, health care would consume nearly 20 percent of the GDP. A large portion of these funds goes to pay for prescription drugs, which has made pharmaceutical companies members of the most profitable industry in the United States. It seems that the FDA is an integral part of the societal mechanism that has allowed this to happen, and with the complicity of the FDA, pharmaceuticals can get on with the business of the drugging of America—and the rest of the world.

Cranial Electrotherapy Stimulation has even been proven to be effective in the recovery from drug addiction, but I doubt that it is significantly beneficial to most people in benzodiazepine withdrawal. While some benzo people have told me anecdotally that they have found CES to be somewhat helpful, I used two different CES devices over a period of months and could find no discernable benefit. The length and quality of my sleep did improve over that period, but the improvement may well have been simply a function of my body recovering from exposure to benzodiazepine on its own. CES may alleviate some of the symptoms of withdrawal from other drugs but the GABA down-regulation of benzodiazepine withdrawal is unique and there is no indication that Cranial Electrotherapy Stimulation has any effect upon GABA-α receptors—unfortunately.

Chapter Twenty-Two
The Last Dose

In the aftermath of my experience with exaggerated Seasonal Affective Disorder, I continued to use the light box daily until spring was well established. Continuing to use it beyond that time would almost certainly have had a positive influence upon my mood, but it had proven itself so valuable in countering the effects of the changing light in the autumn that I wanted to reserve it for that purpose if I needed it the following autumn. So, during the spring and summer, I contented myself with natural sunlight and made certain that I would often look up into the sky to get light into my body through my eyes.

During this time I now reduced my dosage of Valium by 0.25 mg every two weeks rather than weekly. Even decrementing by that small an amount often caused me to 'feel the cuts.' Withdrawal symptoms were difficult, but could not compare to the irrational fear and suicidal depression of the previous fall. Since they never reached that level of severity, I found it relatively easy to apply my doctrine of not taking withdrawal symptoms personally, not identifying myself with what I was experiencing. As tough as things got, I always relied upon my knowledge that all of my discomfort was caused by a problem with my body's ability to utilize GABA effectively. That problem would be solved within my body eventually, and until then, my sole task was to endure the passage of time.

The months passed, winter gave way to springtime, which gave way to summer . . . and I eventually found myself poised at the final step of discontinuing benzodiazepine, getting from the last 1 mg dosage down to zero. I could hardly believe I had made it to that point, after having spent over two years in the endeavor, the most challenging ordeal of my life. My sanity had depended upon my having been able to maintain an objective, dispassionate perspective from which both to observe and to conduct myself during the process—but there had been numerous times where my sanity had been tested, almost to its limit. A nearly inhuman amount of patience was required and I was continuously amazed at the restraint and courage shown by other people tapering with Valium at the *benzo.org.uk* bulletin board. I had been fortunate in that I could survive on a minimal amount of income, while others I knew online had full-time jobs. I couldn't imagine how they could work forty hours a week while coping with withdrawal from benzodiazepine. Also, I had a supportive spouse, when I knew that other people had to endure the trial either alone, or—worse—with unsympathetic, sometimes even resentful, family members.

I prepared to implement a plan I had devised for the final stage of tapering, a *liquid dilution method*. It was apparent that a reduction by too great a percentage of the previous dose caused the body to be in sudden want of its accustomed amount of benzodiazepine, resulting in distress. The distress often caused an increase in the number and severity of withdrawal symptoms, which itself was a form of distress. As I had approached 1 mg in dosage, I began to feel each cut quite significantly, and usually spent five to seven days in some sort of reaction before I stabilized. That was because as the overall dosage got lower, the percentage of each cut to the total dosage increased. To avoid this effect, I then decided to use a dilution method for getting from 1 mg diazepam down to 0 mg.

Reducing by decrements of 0.25 mg of Valium seemed initially quite small, but as the overall dose lowered, the percentage 0.25 mg constitutes of the total dose increased exponentially. At a dose of 4 mg, a cut of 0.25 mg represents 6.25%, a percentage small enough that it doesn't take much for the body to adjust to it. At 2 mg, that same cut of 0.25 mg becomes a reduction of 12.5%. At 1 mg, it becomes fully 25% of the total dose. Had I continued to use 0.25 mg cuts, the final reductions from 1 mg down to 0 would have been 25%, then 33%, then 50%, and, finally, 100% of the total dose.

I decided that it would suit me better to use a continually variable reduction method, where the percentage of each cut actually decreased—with the exception of the final dose, where stopping will always represent 100% of the previous amount.

I was two months from completing the taper, so I calculated the entire quantity of diazepam I would consume over the course of those sixty days. I then had my pharmacy compound that amount of diazepam and put it into a liquid suspension where 1 milliliter (ml) of liquid equaled the amount of diazepam in a 1 mg tablet. This liquid came in a bottle, and I also got a bottle of just the plain suspension medium, *i.e.,* the liquid without any diazepam in it.

The way I took a dose was actually quite simple. Using an oral syringe (without a needle) I would remove from the medicine bottle 1 ml. from the liquid containing diazepam and ingest it by mouth. Then I put 1 ml. of the plain suspension liquid (containing *no* diazepam) back in that bottle, so it constantly had the same total amount of liquid in it. In this way, the amount of diazepam was constantly being diluted. I was making a 'cut' with each and every dose I took. For example, the first day of taking the liquid, I ingested 1 mg of diazepam. On the second day, because I had diluted the mixture when I replaced the liquid in the bottle, the dose of one ml. only contained 0.9667 mg of diazepam. On the following day, one ml of liquid contained 0.9344 mg of diazepam, and so on.

The amount by which a decrease was made got smaller and smaller all the time, and each day's 'cut' represented a continually smaller percentage of the previous dose. The final cut was the equivalent of a mere 1/5000th of a milligram of diazepam, to a final dose of 0.1353 mg of diazepam. The last dose from which someone ends their taper is called 'jumping off' by people discontinuing Valium. I doubt anybody ever 'jumped off' from a lower amount than 0.1353 mg. Perhaps I was being overly cautious by using this method, but I reasoned that it would be better for me to be too cautious than not cautious enough. I did note that after I switched to liquid dilution dosing, the more neurological symptoms such as fasciculations and Restless Leg Syndrome disappeared. Other symptoms persisted, but that was to be expected: even though this was a more gentle method of reducing the dosage, it was a reduction nonetheless, and my central nervous system was bound to react to it.

It was a momentous day for me when I took my last dose of Valium. I was immensely proud of my achievement. But I felt both exhilarated and somewhat apprehensive. *What would happen now?* I knew that, unlike any other substance of addiction, the problem I still faced was not intoxication by the drug itself, but my body's need to recover its ability to utilize GABA effectively. No longer having benzodiazepine present in significant amounts would at least mark an end to my ingesting a material that actually caused down-regulation of GABA. But how quickly would my body be able to correct the condition? I knew from *The Ashton Manual* that substantial recovery is usually expected to occur between 6 and 18 months after the final dose of benzodiazepine is taken. I had also, however, read many an account of people who, weeks or months after having taken their last benzo, found themselves with withdrawal symptoms far worse than anything they had encountered during the tapering process. What, I wondered, was going to happen to me?

I had left *benzo.org.uk's* bulletin board just before I switched to the liquid dilution method. The place had been a virtual home to me during the years of tapering, but I decided that it was time to focus all of my attention on my recovery, and on a more active life away from my computer, now that my condition was tolerable enough to handle it. I had made good friends with so many people at the website, but it was time to shift gears, mentally and emotionally. My first days off benzos were rather tentative—I kept checking to see if anything, either benign or untoward, was happening to me. Fortunately, I had no reaction to the last dose, but that was to be expected. After all, I had tapered down to such a miniscule amount of benzodiazepine, and since its formulation was diazepam, it took a few days after the final dose for all of the drug itself to be assimilated by my body. Further, I knew that benzodiazepine insinuates itself into fat cells, so it would be a considerable amount of time before my body had scavenged the last of the benzo molecules from my tissues.

The immediate post-benzo period turned out to be anticlimactic.

During the ensuing nine months, I continued to play host to nearly all of the withdrawal symptoms I had had during the tapering process, most notably, sleep problems, fatigue, and periodic returns of muscle twitches and spasms in my legs, intermittent numbness of my lips, and the perennial dysthymia and anhedonia. The severity of these phenomena diminished so slowly that I could never tell that my circumstances were improving other than by thinking back to how I had been feeling a month or so previously. Occasionally, I would feel considerably better, but that would only last for a few days at a time. I had read the accounts of people who began to feel actually good as soon as they had finished taking the drug. That was evidently not going to happen to me. It would appear that for some, benzodiazepine is toxic simply as a chemical. Once it is no longer present, their bodies return rather quickly to a functional state. For that type of person, my slow and gentle schedule wouldn't be appropriate; unfortunately, however, there is no way of determining in advance whether any individual is prone toward benzodiazepine toxicity or not.

I had anticipated that getting off the drug would feel sort of like emerging from a cocoon, but it hardly resembled so dramatic a change. I still had to manage my neurological symptoms, and I still could never count on whether I was going to have an okay day or not, other than by experiencing it. People whose health is normal have the luxury of taking it for granted, of being fixated on other aspects of their lives. For me, I continued to be focused on my condition every day, an unnatural occupation for a human being. As the months went by, however, I became incrementally more capable of fulfilling my role as a husband and father. I was still prone toward that phenomenon familiar to people with any of the various forms of Chronic Fatigue Syndrome, where any unusual exertion results in a day or more of profound exhaustion, so I continued to manage myself rather carefully. I kept waiting for the heralded 'windows' to show up, but they never did. That was okay, I accepted whatever happened to me. The specific characteristics of both withdrawal and recovery from benzodiazepine are unique to each individual who goes through them, and I was happy that, if nothing else, I had gotten myself safely off of the drug and was now in the process of reclaiming my life, such as it was. I was only too aware of how fortunate I was to be free of having to take every day a drug I despised, simply to maintain a chemical addiction for which I would never knowingly have put myself at risk.

Chapter Twenty-Three
"Ask your doctor"

For those who are currently taking tranquilizers, learning of the existence of a benzodiazepine withdrawal syndrome may lead quite reasonably to the question of whether or not you should stop taking a drug that may pose such a considerable threat to your well-being. After all, you may be at risk of having significant withdrawal effects such as I and numerous other people have had to endure. Then again, your body may have a minimum of difficulty in reestablishing the effective use of its GABA, and therefore withdrawal from benzodiazepine may well prove to be a straightforward procedure with few problems, if any. Or, you may find yourself somewhere in between the extremes, where withdrawal results in some degree of discomfort which impacts your quality of life. Unfortunately, there is no way to know in advance what any individual's reaction might be. Even people who have been successful in getting off of benzodiazepine in the past may find themselves in profound difficulty in subsequent attempts to discontinue the drug.

It is an unfortunately common feature of modern life that attempts to inform or to market often use the alarmist technique of inspiring fear in people; fear is, evidently, an effective motivator in getting people to take action. As such, it is decidedly not my intention to frighten the readers of this book into believing that everyone who takes the drug is in grave danger from benzodiazepine withdrawal. It is my responsibility as a communicator, however, to share information accurately and truthfully—as I perceive it. I spent three years in continual contact with people having extraordinary hardships as a result of trying to come off benzodiazepine. I can only conclude that since some of those currently on any of the benzodiazepine drugs are likely to have a similar experience, I should provide as much relevant data as possible and let readers cull from it what they find useful to their particular circumstances. In the interest of being thorough, I should add a reminder that I am neither a doctor nor a medical professional, and the opinions and ideas expressed in this book should therefore not be considered as medical advice. They are offered for informational purposes only, the better to empower the informed consent of the readers.

How likely is it that you might face exceptional difficulties in trying to discontinue benzodiazepine? Unfortunately, there is no accurate way of answering that question because there are so few definitive data on benzodiazepine withdrawal. Thorough drug studies are usually only made *before* a drug is introduced to the public. The follow-through on what happens to consumers afterward is sketchy, and consists mainly of post-marketing reports by the drug manufacturers themselves. And thus, drugs that gained approval because of studies based upon use by mere tens or hundreds of human test subjects go into commercial use where they are used by millions of people with little oversight. Benzodiazepines have been used by *hundreds* of millions of people—and yet, there is little real science about what happens to those who attempt to discontinue them.

In *Detoxification from Benzodiazepines: Schedules and Strategies* published in the *Journal of Substance Abuse Treatment* in 1991, Drs. Paul Perry, Bruce Alexander, and Brian C. Lund suggest that studies show a low incidence of benzodiazepine withdrawal syndrome:

> *The correlation of BZD dose and duration of use to the incidence and severity of BZD withdrawal remains to be precisely quantified. However, these two factors appear to be primary considerations in predicting the likelihood of withdrawal. A review of the controlled studies of long-term therapeutic doses of BZD indicates that nearly 50% of patients ingesting a BZD for an average of three years will experience a minor withdrawal syndrome when the drug is discontinued. Eight months or more of daily use may produce a 43% withdrawal incidence, while use less than 8 months significantly reduces the incidence of withdrawal.*

Various benzodiazepine awareness groups, however, suggest that possibly 50%–80% of benzo users will experience withdrawal symptoms when coming off the drug. Their data appear to be supported by Magellan Behavioral Health/Blue Cross Blue Shield of North Carolina which states in their *Guideline for the Diagnosis and Management of Generalized Anxiety Disorder for Primary Care Physicians:*

> *A large number (40–80%) of patients treated with [benzodiazepines] for 4 months or more develop tolerance and can have a discontinuation or withdrawal syndrome. Patients may experience rebound or withdrawal symptoms, which include tremor, anxiety, agitation, and dysphoria.*

A chart included in a 1990 brochure by Roche Products (UK), Ltd., shows that after one year, fully 25% of people attempting to discontinue Xanax use will be unsuccessful, presumably because the withdrawal symptoms were so daunting they reinstated it. And yet, there is something that doesn't make sense about all of this. If, indeed, hundreds of millions of people have taken benzos over all these years, and a percentage of those who have discontinued its use suffered with benzodiazepine withdrawal syndrome, why is the problem not more widely known? Wouldn't it be statistically likely that each of us would know someone who, like me, had had a life-interrupting reaction to discontinuation?

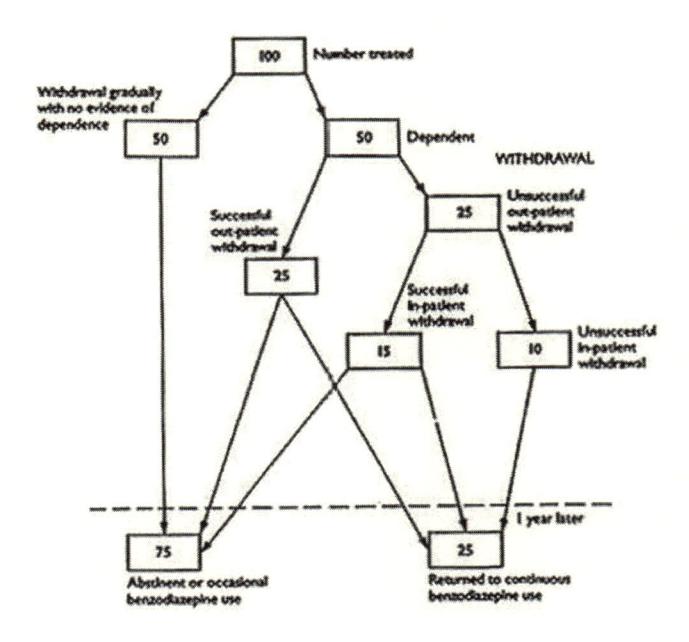

FIG. 8 : *From a brochure by Roche Products (UK), Ltd., sent to physicians upon request*

One possible explanation may lie in the way that such a reaction is dealt with medically. It is quite possible that, since no doctor would imagine that a benzodiazepine could possibly be responsible for the exaggerated problems that benzo withdrawal often presents, doctors would probably treat the withdrawal symptoms medically, *i.e.,* by prescribing medicines. Remember that the idea that a substance can still affect someone long after it is no longer in his or her body would appear illogical and ridiculous. Therefore, the presentation of the often-peculiar symptoms of down-regulated GABA is probably usually misdiagnosed as the emergence of a patient's underlying pathology and the drugs given to ameliorate those symptoms would tend to obscure that the true underlying cause is benzo discontinuation.

I have a friend, whom I will refer to as 'Rachel.' Like me, she had been given the antidepressant, Effexor; like me, her physical reaction to it was characterized as 'agitation,' and, also like me, her doctor's response was to prescribe Xanax. Within a year, Rachel's Xanax use had escalated out of control. Her doctor arranged for her to be admitted to a detoxification facility, where she was diagnosed as having Bipolar Affective Disorder. She was taken off Xanax and Effexor and put on Klonopin and Depakote, an anti-seizure medicine approved for treating epilepsy and severe mania. She was released after two weeks of treatment.

After Rachel got home from detox, she was somewhat shellshocked by the experience, as well as her own manic behavior leading up to it. Depakote has some rather nasty side effects. In addition to presenting a risk for causing serious liver damage and life-threatening pancreatitis, users report harsh side effects: weight gain, exhaustion, confusion, loss of libido, heartburn, and gastrointestinal disturbance. On her own, Rachel tapered herself off of Depakote to avoid the side effects, and found, as she had expected, that she was not manic, did not have Bipolar Affective Disorder. A little research confirmed what she had suspected, that the SSRI antidepressant, Effexor, had a reputation for causing hypomania and, in some, full-blown rapid-cycling mania. The combination of Effexor and Xanax had proved to be her undoing, but while she had been on those drugs, she was incapable of seeing her own situation with enough mental clarity to determine that the medication wasn't suited to her.

After she discontinued the Depakote, her new doctor switched Rachel from Klonopin to a lower dose of Ativan. Life appeared to become increasingly stressful for her, and Rachel found herself—for the first time in her life—using alcohol. She became a 'secret drinker,' but it was clear to her friends from her behavior that she was having a problem with alcohol. After about a year, concerned about her 'alcoholism,' her doctor told her he couldn't treat her anymore and referred her to a local substance abuse program. Using an uncompromising 'tough love' approach, her counselor instituted a three-week taper from Ativan.

After about five days off Ativan, Rachel called me and asked for help. When I went to see her, she was in bad shape. I took her to see my own doctor, and he recognized that she was in severe benzodiazepine withdrawal. He reinstated her on Valium and began an Ashton taper based upon the one he had supervised for me. Within a few days, Rachel was functional again, able to sleep at night and hold down a job.

In going over her history with her, it finally became clear that what had happened was that her rapid detoxification from Xanax had been ineffective: she was still very much benzodiazepine-dependent, and neither the subsequent Klonopin or Ativan had been at a high enough dose to quell the withdrawals. She had developed the classic symptoms of insomnia, nervousness and agoraphobia. Rachel's use of alcohol appeared to be an unconscious way of potentiating the benzodiazepine in the Ativan, but it provided little additional relief and created problems of its own, as alcohol so often does. Once she was established on a Valium taper, Rachel never again drank. Neither the detox facility nor her doctor had recognized benzodiazepine withdrawal syndrome; it wasn't until a correct diagnosis had been made that Rachel was able to function.

I cite this as an example of someone who 'fell through the cracks' in the medical system. Perhaps the reason that benzodiazepine withdrawal syndrome isn't more widely reported is because those who have it get diagnosed and then treated for mental problems. Subsequent drugging would then result in someone actually becoming a 'mental patient,' fulfilling the prophecy of the erroneous diagnosis. I know that if I had not discovered that Xanax was the cause of my own debilitation, I would eventually have checked myself into a hospital to find some relief. I'm sure I would have been diagnosed and drugged until I was somewhat comfortable, but those drugs would have masked the down-regulation of GABA that was the actual source of my invalidism. The effect upon me psychologically of being in a hospital setting, drugged insensate every day, would have led, I'm sure, to a sort of mental death.

The decision to stop taking any medication is not one that should be made lightly. In the case of benzos, that decision is made more complex and difficult for two reasons: first, because the single most crucial aspect of it—the down-regulation of GABA—is so largely unknown in both the medical community and among benzo consumers, and second, because to discontinue use of the drug may or may not lead to benzodiazepine withdrawal phenomena. Medical decisions are invariably linked with the phrase, "Consult your physician." Usually, that is vital advice but with benzos, the possibility that your physician would have any understanding of the possible GABA issue or the withdrawal syndrome it may create is highly unlikely. Without that crucial knowledge, the advice and counsel of your physician might turn out to be unintentionally damaging.

The "ask your doctor" mindset is currently so pervasive that it almost implies that it is *your doctor*, a trained professional, who should be the one to decide what drugs you should be on, not *you*. This exaggerated reliance on the judgment of doctors has led to the modern phenomenon of polydrug dependency, where we are to take an ever-expanding multiplicity of maintenance medications daily, with some drugs being given to counteract the side effects of other drugs. Such a practice may lead to functionality, but that is not at all the same thing as 'health.'

We seem to have relinquished the responsibility for ourselves, and that is a dangerous precedent for free people in a free society. These are *our* bodies, and we are ultimately responsible for them. The power to choose what medicines we take should repose with us; the role of the doctor should be to help us make informed decisions, not to make those decisions for us. If you are taking benzodiazepine, and have taken it for more than two to four weeks, the doctor who prescribed it for that length of time may have already put you in harm's way. Would it be wise to ask such a person, who has already disregarded the danger of drug dependency, about whether it would be appropriate to discontinue benzodiazepine?

In my case, the conclusion to stop taking Xanax was not reached by taking a rational look at the drug I was taking and assessing its risks. My decision occurred because I had already gone into a damaged state as a result of having reached tolerance. Once I realized that it was a drug that had caused my condition, I immediately took steps to stop taking it. It is only due to the problematic nature of discontinuing benzodiazepine that it took a little over two years to get off of it safely. If you are currently taking a benzo and are concerned, you may want to ask yourself two questions, *"Why would I want to continue taking it?"* and *"Why would I want to stop taking it?"* Your decision would follow a careful examination of the pros and cons revealed by the answers to those two questions. Before even beginning such an inquiry, however, it is imperative to keep in mind that you should **never, never, never stop taking any benzodiazepine drug abruptly!** To do so may lead to potentially fatal convulsions and seizures, and may bring on severe benzodiazepine withdrawal effects which could persist for an extended period of time.

"Why would I want to continue taking a benzodiazepine?"

Benzodiazepines are remarkably effective drugs. One of their chief virtues is that they can be fast-acting, providing quick relief. Also, compared to earlier barbiturate-type sedatives, benzodiazepines present fewer pronounced side effects such as lethargy or slurring of speech at therapeutic doses. As such, when taken for insomnia, they may produce less of a 'hangover' the next day; taken for anxiety disorders, patients may feel less 'sedated' and more like their normal selves.

People whose use of a benzodiazepine has allowed them to become more functional would be quite understandably reluctant to stop using them: a person with insomnia who can now depend upon being able to get to sleep on time by taking a benzo such as Dalmane or Restoril, someone with Social Anxiety Disorder (SAD) whose use of Klonopin helps them feel at ease in the company of others, or someone whose Panic Disorder would keep them housebound without the ability to take Xanax at the first sign of an incipient anxiety attack. To such individuals, benzodiazepines constitute what may be, to them, a medical tool of inestimable value. The dilemma lies in the illusion that the aid the drug provides will continue to work for them forever, without creating problems of its own.

"Why would I want to stop taking a benzodiazepine?"

In answer to that question, Prof. C. Heather Ashton says, in *The Ashton Manual*:

> *[L]ong-term use of benzodiazepines can give rise to many unwanted effects, including poor memory and cognition, emotional blunting, depression, increasing anxiety, physical symptoms and dependence. All benzodiazepines can produce these effects whether taken as sleeping pills or anti-anxiety drugs. . . .*
>
> *Furthermore, the evidence suggests that benzodiazepines are no longer effective after a few weeks or months of regular use. They lose much of their efficacy because of the development of tolerance. When tolerance develops, 'withdrawal' symptoms can appear even though the user continues to take the drug. Thus the symptoms suffered by many long-term users are a mixture of adverse effects of the drugs and 'withdrawal' effects due to tolerance. The Committee on Safety of Medicines and the Royal College of Psychiatrists in the UK concluded in various statements (1988 and 1992) that benzodiazepines are unsuitable for long-term use and that they should in general be prescribed for periods of 2–4 weeks only.*
>
> *In addition, clinical experience shows that most long-term benzodiazepine users actually feel better after coming off the drugs.*

But Prof. Ashton adds, however: *"The advantages of discontinuing benzodiazepines do not necessarily mean that every long-term user should withdraw. . . . The option is up to you."*

The option, the choice, and therefore, the decision, is one that any user of benzodiazepine should consider, and consider most carefully and with great deliberation. While it is possible to sustain benzo use for very long periods, there is an inherent danger that, at any time, side effects from or tolerance to the drug may cause a cascade effect of disabling conditions. That is precisely what happened to me; it caused me not merely to lose three years of my life but to have spent those years in what usually felt like a living hell. It would have been far preferable had I gotten myself off the drug long before its use led to debilitating problems and extremes of discomfort.

There is some anecdotal evidence to suggest that antibiotic use might be linked to benzodiazepine withdrawal syndrome. At the online community associated with *benzo.org.uk,* people sometimes mentioned that they had had bad reactions to antibiotics, resulting in a worsening of their withdrawal symptoms. It struck me that, shortly before I began to go into tolerance with Xanax, I had been treated for babesiosis—with a potent antibiotic.

It is not widely recognized, by patients *or* by doctors, that some antibiotics impact the central nervous system. One class of antibiotics, the fluoroquinolones, is particularly noted for its side effects. One of the people with whom I have kept in closest touch, having both finished our tapers at the same time together, had only taken a benzodiazepine, Ativan, for two months when she was prescribed Levaquin, a fluroquinolone antibiotic. After taking but two of the pills she had to stop because her reaction was so severe: overwhelming fatigue, "black-hole depression" and debilitating pain in her jaw joint which persisted for weeks. Even that brief exposure to fluoroquinolone was enough to cause her to go into a state of tolerance with Ativan, thus precipitating her long ordeal with benzodiazepine withdrawal syndrome.

Many doctors appear to remain in denial that the drugs they prescribe can have such profound effects. The North Carolina Board of Pharmacy created a policy in 1992 requiring that pharmacy managers report any deaths related to drug use. Alicia Cullen, a Pharm.D. Candidate at the University of Arkansas for Medical Sciences, and David R. Work, the Executive Director of the North Carolina Board of Pharmacy, reviewed the data collected between 1992 and 2001. They found that 10.5% of drug-related deaths were a result of schedule II controlled substances, the drugs with a high potential for abuse which merits their intense scrutiny by law-enforcement agencies. However, an almost equal number of deaths, 9%, was caused by the fluoroquinolone antibiotics, *Ciprofloxacin, Gatifloxacin, Levaquin, Levofloxacin, Ofloxacin, Omniflox,* and *Trovafloxacin*. In other words, these drugs, whose use is not especially monitored, caused almost the same amount of mortality as drugs notorious for their danger.

Fluoroquinolones have been associated with one of the more bizarre—and devastating—side effects ever reported. The package insert for Levofloxacin contains, among admonitions about other potential side effects, these warnings:

Peripheral Neuropathy
Rare cases of sensory or sensorimotor axonal polyneuropathy
affecting small and/or large axons resulting in paresthesias, hypo-
esthesias, dysesthesias and weakness have been reported in patients
receiving quinolones.

Quinolones may cause central nervous system (CNS) events
including nervousness, agitation, insomnia, anxiety, nightmares,
or paranoia.

However, there is also this warning:
> *Ruptures of the shoulder, hand, and Achilles tendons that required surgical repair or resulted in prolonged disability have been reported in patients receiving quinolones. Tendon rupture can occur during or after therapy with quinolones.*

Evidently, as a direct result of taking a fluoroquinolone antibiotic, tendons can become so inflamed that they spontaneously tear away from the bone. So severe is the damage that it often can only be repaired by surgical intervention. In cases where actual rupture doesn't occur, the swelling in tendons, muscles or joints can be so painful to patients that they become disabled, and the condition may persist for many months.

Many studies have been done on this phenomenon. The opening sentence in one such study, *How Does Levofloxacin Compare to Other Antibiotics?*, by Marcia L. Brackbill, Pharm.D., and Connie L. Barnes, Pharm.D., Director, Drug Information Center Campbell University School of Pharmacy, states that, *"Fluoroquinolone antibiotics such as ciprofloxacin, norfloxacin and ofloxacin are commonly prescribed due to their broad spectrum of antimicrobial activity and relatively safe adverse effect profile."* Another study, by Richard M. Harrell, entitled, *Fluoroquinolone-Induced Tendinopathy: What Do We Know?*, begins with the sentence, *"Fluoroquinolones are relatively safe, effective antibiotics."* In the Issue 14 of the 2001 *Journal of Nephrology*, Moreno Malaguti, Luigi Triolo and Marco Biagini of the Department of Nephrology, in San Paolo Hospital, Civitavecchia, Italy, begin their study, *Ciprofloxacin-associated Achilles tendon rupture in a hemodialysis patient*, with an almost identical sentence: *"Fluoroquinolones (FQ) are relatively safe and effective antibacterial agents."* Numerous other articles employ that same phrase, further on in their texts. Relatively safe? Relative to what? The 2001 report by the North Carolina Board of Pharmacy shows clearly that these drugs are as dangerous as the most dangerous drugs on the market.

Due to the potential egregious risks inherent in these antibiotics, as well as in an effort to prevent drug-resistant disease strains due to overexposure to them, their best use has been determined to be as a 'second line' defense, *i.e.,* after an older antibiotic has been employed but failed to meet treatment goals. Studies have shown, however, that both hospitals and doctors in private practice routinely employ them as their inceptive medical response. Issue 163 of the 2003 *Archives of Internal Medicine* featured a study of one hundred consecutive emergency room patients treated with a fluoroquinolone antibiotic. The study reported:

Of 100 total patients, 81 received an FQ [fluoroquinolone] for an inappropriate indication. Of these cases, 43 (53%) were judged inappropriate because another agent was considered first line, 27 (33%) because there was no evidence of infection based on the documented evaluation, and 11 (14%) because of inability to assess the need for antimicrobial therapy. . . . Of the 19 patients who received an FQ for an appropriate indication, only 1 received both the correct dose and duration of therapy.

Inappropriate and injudicious prescription of this class of antibiotics continues, even in spite of the fact that in 2004, the FDA upgraded the warnings on required package inserts. Perhaps the perception that such drugs are 'relatively safe' comes from the idea that serious side effects are 'rare' because they occur in less than one percent of patients. Viewing the patient population as a whole, however, the scale of potential victims is considerable. According to a Knight Ridder analysis of data for the year between July 2002 and July 2003, there were 12.2 million prescriptions written for just one fluoroquinolone, Johnson & Johnson's Levaquin. If one tenth of one percent of those prescriptions resulted in a serious adverse reaction, that would mean 1,220 people would be injured, or possibly killed, by the drug. Though they represent only a small percentage of the total consumer/patient base taking Levaquin, that is a significantly large number of individuals put at risk, perhaps inappropriately.

While the danger of fluoroquinolone-caused tendinitis and rupture is obviously not to be minimized, of particular interest to people taking benzodiazepine or contending with benzodiazepine withdrawal syndrome are the CNS effects, *"nervousness, agitation, insomnia, anxiety, nightmares, or paranoia"* and "peripheral neuropathies" referred to in the FDA warnings as, *"pain, burning, tingling, numbness and or weakness... deficits in light touch, pain, temperature, position sense, vibratory sensation, and or motor strength."* Such side effects of the fluoroquinolones are widely reported as benzodiazepine withdrawal symptoms. This would suggest that the very same neurological functions affected by the one drug are affected by the other, as well.

That theory is supported by research. An article entitled, *Possible interaction of fluoroquinolones with the benzodiazepine-GABA-α receptor complex,* published in Issue 30 of the 1990 *British Journal of Clinical Pharmacology,* states that, *"There is experimental evidence that CNS adverse effects of different quinolones could be related to an inhibitory effect on GABA-binding with a subsequent inhibition of GABA-mediated inhibitory transmission."* Fluoroquinolone antibiotics evidently reduce the ability of the GABA-α receptor sites to bind GABA, which hampers the ability to calm neural cells from excitatory states. Since that is the same problem Professor Ashton identified as the cause of benzodiazpine withdrawal syndrome, it is no wonder the symptoms of Central Nervous System distress are identical between adverse reactions to fluroquinolones and benzodiazepines. For people who are currently taking benzos, or worse, who are in a state of benzodiazepine tolerance or withdrawal, the introduction of a fluoroquinolone would be highly likely to disturb an already compromised relationship between GABA and GABA-α receptors. Like benzodiazepines, the effects of fluoroquinolones can extend far past the time of termination of taking the drug, which suggests that they, too, may cause down-regulation of the GABA-α receptors. As such, it would be highly advisable for people currently struggling with benzo issues to preclude, if at all possible, taking fluoroquinolone antibiotics. To avoid perhaps difficult and potentially unproductive discussions with doctors, my own strategy is simply to state that I am allergic to fluoroquinolones, leaving it to the doctors to provide an alternative antibiotic therapy.

With all the apparent dangers of prescription medicines, it might seem the wisest course would be simply never to take any. That is a too simplistic approach to what is a far more complex situation. Drugs themselves are morally neutral, neither good nor bad. It would be best to view them dispassionately, with perhaps equal measures of hope and skepticism: 'hope' that a given drug may help alleviate a medical condition, and 'skepticism' that it will do so without causing other harm. Since they can be potentially hazardous, and those hazards may very well not be acknowledged by health care providers, prescription drugs should be approached with great caution.

Chapter Twenty-Four
Addiction *vs.* Dependence

Trying to make an 'informed decision' about whether to continue taking benzodiazepines is difficult if the information you consider is conflicting and contradictory. In the Information Age, many people seek out answers on the Internet. The Pew Internet & American Life Project's November 2004 survey found that fully 40% of people using the Internet had searched for information about prescription or over-the-counter drugs. It is often unclear how useful—or of what quality—the available information is. If someone were to use 'social anxiety disorder benzodiazepine' on Google, some 309,000 websites would be returned as relevant. Clicking on the links among the top five websites sought out under those criteria, yielded a wide sample of opinions.

From *www.socialanxiety.factsforhealth.org:*
> *Clonazepam (Klonopin) is the most extensively studied benzodiazepine for social anxiety disorder and has been shown to have significant beneficial effects. Benzodiazepines have the advantage of decreasing anxiety faster than the other medications. Benzodiazepines have the disadvantage of not treating depression and long-term use can cause physical dependency.*

From *www.socialfear.com,* a website devoted primarily to Social Phobia (SP):
> *Klonopin (clonazepam): Klonopin is extremely effective for SP and usually works great. Klonopin can be taken either 'as needed' or everyday. 'As needed' (prn) use can be done up to twice per week, and will usually provide excellent effect within 30 minutes, lasting several hours to 1 day (typical dose .25–.75mg). Taken 'long term', Klonopin may be used alone, although sometimes a non-sedating antidepressant is added if depression also exists.*
>
> *Myths About Benzodiazepines:*
> * *'Benzodiazepines are addictive': FALSE for non drug addicts with anxiety disorders.*
> * *'Benzodiazepines are hard to quit': FALSE (for SP, not GAD). Taper slow. Can cross-taper gabapentin if desired.*
> * *'Benzodiazepine dose keeps escalating': FALSE. Dose stabilizes after a few months with continued efficacy.*
>
> *Possible Drawbacks of Long Term Benzodiazepine Use:*
> * *Depression may be aggrevated[sic].*
> * *Reduced mental sharpness may occur.*
> * *Reduced motivation may occur.*

From *www.npadnews.com*, National Panic and Anxiety Disorder News:

> *Symptoms upon tapering. Studies indicate that between 35 and 45 percent of patients are able to withdraw from the BZs without difficulty. Of the others, three different problems can arise. These are symptoms of withdrawal, rebound, and relapse, which can sometimes occur simultaneously.*
>
> *Dependence and withdrawal symptoms.*
> *Physical dependence means that when a person stops taking a drug or reduces the dose quickly, he or she will experience symptoms of withdrawal. BZ withdrawal symptoms usually begin soon after reduction of the drug begins. . . . These symptoms can be bothersome but are usually mild to moderate, almost never dangerous, and resolve over a week or so.*
>
> *At least 50% of patients experience some withdrawal symptoms when they stop taking a benzodiazepine, and almost all patients experience strong withdrawal symptoms if they stop the medication suddenly. Most experts now taper quite slowly, often taking months to completely discontinue the benzodiazepine.*
>
> *Panic patients seem to be more susceptible to withdrawal symptoms than those with other anxiety disorders.*
>
> *Between 10 to 35 percent of patients will experience the rebound of anxiety symptoms, especially panic attacks, when they discontinue the BZs too rapidly.*

From *www.aafp.org*, the American Academy of Family Physicians:
The benzodiazepines are fast-acting, well-tolerated anxiolytics that have shown efficacy in the acute treatment of social phobia, but they have also revealed some significant drawbacks related primarily to difficulties with discontinuation. Controlled studies of alprazolam (Xanax) and clonazepam (Klonopin) report acute-treatment improvement rates ranging from approximately 40 to 80 percent, with clonazepam showing more favorable results.

However, standing dosages are sometimes difficult for patients to taper and discontinue without symptomatic worsening and a high risk of acute relapse.

Because of their ability to produce physical dependence, benzodiazepines must be used with caution in patients with a history of substance abuse, a condition often associated with social phobia.

... Use of benzodiazepines in therapeutic dosages does not lead to abuse, and addiction is rare.... All benzodiazepine therapy can lead to dependence; that is, withdrawal symptoms occur once the medications are discontinued. Withdrawal symptoms include anxiety, irritability and insomnia, and it can be difficult to differentiate between withdrawal symptoms and the recurrence of anxiety.... Patients may experience rebound anxiety (akin to the rebound hypertension that occurs when some antihypertensives are discontinued) once the tapering process is completed, but this is transient and ends within 48 to 72 hours. Once the rebound anxiety ends, a patient may re-experience the original symptoms of anxiety, referred to as recurrent anxiety.

> *Although few controlled studies support the long-term use of benzodiazepines, GAD [Generalized Anxiety Disorder] is a chronic disorder, and some patients will require benzodiazepine therapy for months to years. Generally, patients who present with acute anxiety or those with chronic anxiety who undergo a new stressor ('double anxiety') should receive benzodiazepine therapy for several weeks. Patients may be less tolerant of anxiety that recurs when the benzodiazepine is discontinued and, if necessary, it may have to be resumed indefinitely. Patients who use benzodiazepines chronically tend to be elderly, to be in psychologic[sic] distress and to have multiple medical problems.*

From the 'Anxiety Community' at *www.healthyplace.com*:
> *High-potency benzodiazepines relieve symptoms quickly and have few side effects, although drowsiness can be a problem. Because people can develop a tolerance to them, and would have to continue increasing the dosage to get the same effect, benzodiazepines are generally prescribed for short periods of time.*

> *One exception is panic disorder, for which they may be used for 6 months to a year. People who have had problems with drug or alcohol abuse are not usually good candidates for these medications because they may become dependent on them.*

From the University of Maryland Medical Center's website at *www.umm.edu*:
> *Benzodiazepines are effective medications for most anxiety disorders and have been the standard of treatment for years. However, their use has been associated with a risk for dependency and abuse, and so they have been supplanted in many cases by SSRIs*

and by newer antidepressants. Benzodiazepines include the following:

• *Alprazolam (Xanax) and clonazepam (Klonopin) are effective for panic disorder, some phobias, and generalized anxiety disorder.*
• *Other benzodiazepines, including diazepam (Valium), lorazepam (Ativan), and chlordiazepoxide (Librium), are used mainly for generalized anxiety.*

Loss of Effectiveness and Dependence. Eventually these drugs may lose their effectiveness with continued use at the same dosage. As a result, patients may need to increase their dosage to prevent anxiety. Patients then can become dependent on these agents; this can occur after as short a time as three months. It should be noted, however, that patients with generalized anxiety disorder rarely become tolerant to their effects. Evidence also suggests that the risk for abuse exists only in people who are already susceptible to substance abuse.

Withdrawal Symptoms. People who rapidly discontinue benzodiazepines after taking them for even four weeks can experience rebound symptoms. The longer the agents are taken and the higher their dose, the more severe these symptoms can become.

Deliberating over these bytes of information, we learn from *npadnews.com* that *"At least 50% of patients experience some withdrawal symptoms when they stop taking a benzodiazepine'* and *'Panic patients seem more susceptible to withdrawal symptoms...."* while *socialfear.com* doesn't mention discontinuation problems in its list of *"Possible Drawbacks of Long Term Benzodiazepine Use"* Withdrawal symptoms may be *"bothersome"* according to *npadnews.com*, but *"almost never dangerous, and resolve over a week or so...."* although *aafp.org* says that tapering and discontinuation is sometimes difficult without symptoms getting worse and a *"high risk of acute relapse."*

We are further informed by *aafp.org* that substance abuse is *"often associated with social phobia"* even though *socialfear.org* states that 'Benzodiazepines are hard to quit' only for people with Generalized Anxiety Disorder, not for those with Social Phobia. *aafp.org* goes on to say that some patients *"will require benzodiazepine therapy for months to years"* and those who have problems upon discontinuation that benzodiazepine *"may have to be resumed indefinitely"* in spite of the caution by *socialanxiety.factsforhealth.org* that *"long-term use can cause physical dependency."* It is no wonder that doctors routinely exceed the dosing guidelines for benzodiazepines set by the FDA when it is obviously widely believed that some applications for the drug warrant long-term use even though it appears that difficulty in terminating such long-term benzodiazepine therapy is well established.

A patient looking for cogent advice from these sources would find no clear strategies presenting themselves. In the Information Age, with the glut of data available to us, it takes the application of diligent evaluation while considering information to determine its validity. This very book is, in its way, no more authoritative than any of the chunks of information someone might find on the Internet. Although I have endeavored to research the things I've written as thoroughly as possible, I am neither a scientist, nor a researcher, nor a doctor, and the ideas I have expressed within these pages are, largely, my own perceptions and opinions. My hope is that readers will view what I have written with skepticism and take from it only those ideas that survive their own gauntlet of examination.

The one thing that all of the websites referenced above appear to agree on is that the people who are at risk of 'addiction' to benzos are only those who have a previous history of substance abuse, not those on therapeutic doses. *aafp.org* says that 'addiction is rare' but that 'all benzodiazepine therapy can lead to dependence.' The idea that some people, such as those with Social Anxiety Disorder, or those without a history of drug abuse, *can't* become addicted to benzodiazepine is irrational—how could either circumstance create an immunity to the GABA down-regulation which characterizes benzodiazepine withdrawal syndrome?

An Arizona clinic which offers, among other modalities, Cognitive Behavioral Therapy for people suffering from *Generalized Anxiety Disorder (GAD)* tells prospective clients on their website:

Anti-anxiety agents, such as Ativan and Klonopin: These are typically the agents of choice for starting anxiety management.

Many 'primary care' physicians (GPs) have not been trained in the anxiety disorders and see these medications as being 'addictive'.

However, these medications are NOT addictive for people with clinical anxiety disorders.

Over three dozen research studies report that people with clinical anxiety disorders do not become drug addicts as a result of temporary anti-anxiety use. These medications can be very helpful for people with panic/agoraphobia. Find a psychiatrist who understands this. These medications are tolerated well and almost always help. There are few side effects (e.g., tiredness at first) and they work quickly. There seems to be more research support for the use of Klonopin (clonazepam) in the treatment of anxiety than for the other anti-anxiety medications.

If a professional tells a person with a definable, DSM-IV anxiety disorder that the anti-anxiety agents may prove addictive to them, the professional (a) is not aware of research in the area of anxiety, and (b) should probably not be treating you. The anti-anxiety agents work, they are safe, and people with anxiety disorders usually stay on a low dosage while going through CBT.

These medications are nothing to worry about. When stopping anti-anxiety use, it is necessary to taper off the medication slowly, by reducing the dose over a period of 3 to 4 weeks.

Note its claim that someone with a clinical anxiety disorder cannot become addicted to benzodiazepines, as if the disorder itself can somehow convey immunity to dependency. This website concurs with the other equally authoritative-sounding online sources which state that the only people who are at risk of addiction to benzodiazepines are those who have a previous history of substance abuse. Setting aside other logical inconsistencies, the most glaring error in the reasoning of such an idea is: what if the first substance-of-abuse a potential drug addict is exposed to is benzodiazepine? Wouldn't such a person form an addiction to it, or is it necessary to abuse *other* substances first in order to abuse benzodiazepines? How is it that intelligent, highly educated medical personnel could fail to recognize the erroneous thinking behind such beliefs?

Someone struggling with withdrawal symptoms over a prolonged period may question what the difference is between 'addiction' and 'dependence.'

The free online encyclopedia, *wikipedia.com*, offers some particularly clear definitions:

> *Drug addiction has two components: physical dependency, and psychological dependency. Physical dependency occurs when a drug has been used habitually and the body has become accustomed to its effects. The person must then continue to use the drug in order to feel normal, or its absence will trigger the symptoms of withdrawal. Psychological dependency occurs when a drug has been used habitually and the mind has become emotionally reliant of its effects, either to elicit pleasure or relieve pain, and does not feel capable of functioning without it. Its absence produces intense cravings, which are often brought on or magnified by stress. A dependent person may have either aspects of dependency, but often has both.*

Both descriptions of dependency, physical and psychological, are often present in patients on long-term therapeutic regimens of benzodiazepine use. In the sudden absence of the drug, a patient would almost certainly 'not feel capable of functioning.' Perhaps the defining factor in addiction *vs.* dependence would be that of substance *abuse*, where someone exceeds therapeutic doses of a medication in order to bring about or maintain a chemically-induced state of intoxication or 'escape.' This modality is often characterized by drug-seeking behavior, and secrecy in obtaining and implementing the substance of abuse. Excesses may require illegal activities in order to obtain sufficient quantities of the substance. Medical users of drugs rely upon the amount of the drug prescribed by their doctors, and dosage only escalates as various thresholds of tolerance are reached and exceeded. The incidence of abuse by benzodiazepine users is relatively low, which has perhaps led many in the medical field to conclude that addiction to benzodiazepines is rare. One feature typical in substance abuse is that the ability of addicts to function socially is, often greatly, impaired by their drug-of-choice. By contrast, *use* of a benzo as a therapeutic medicine, almost by definition, is sustained because it *increases* a patient's ability to function normally. The very purpose of the drug in those instances is to alleviate underlying conditions, such as anxiety or insomnia, which would otherwise interfere with usual behaviors, attitudes, and activities.

It was not until some months after I had gotten off benzodiazepine altogether that I had a rather startling realization: during the rigors of tolerance while I was still taking Xanax and during the two years of tapering with Valium *it never once occurred to me to take a higher dosage to relieve my suffering!* I simply never had that thought. The idea that at any moment I could have forestalled, even temporarily, the wretchedness I was experiencing by increasing the amount of benzodiazepine in my system never entered my head. That is rather remarkable in that a constant struggle *not* to increase the intake of the drug-of-choice is characteristic of the addict. As soon as I had realized that Xanax was damaging me, my thoughts centered on getting myself safely off of it, and, evidently, never strayed from that goal no matter how badly I felt. Of course, I had reached an absolute crisis point with Xanax. Users of benzodiazpines who continue to escalate their drug use probably do so simply because any attempt to reduce dosage brings on withdrawal symptoms so horrendous that they cannot be tolerated. Therefore, 'addicts' may increase their drug use in an attempt to feel pleasure, but the motivation in people with iatrogenic habituation is most likely the avoidance of profound distress.

While patients dependent upon benzodiazepines may not be seen as 'abusers' and, therefore, are beyond the moralistic societal judgments with which addicts are regarded, the difference between 'dependency' and 'addiction' becomes one merely of semantics when the rigors of withdrawal begin to occur.

In a briefing paper entitled, *SSRIs & Withdrawal/Dependence*, delivered in June of 2003 by the eminent psychopharmacologist and medical historian, Dr. David Healy, he states:

The idea of therapeutic drug dependence or normal dose dependence re-emerged with the crisis surrounding benzodiazepine dependence in the 1980s. The clinical establishment reacted to the suggestions that the benzodiazepines caused dependence by arguing that there was no tolerance to the benzodiazepines, that these drugs were not abused to any great extent on the street, that the drugs were clearly beneficial in therapeutic situations and as such to talk about the benzodiazepines being addictive was misleading.

From the point of view of the patient however the great concern about the benzodiazepines was that it might not be possible to stop treatment. These drugs led to individuals being hooked in the sense that they were not at liberty to stop.

To someone taking benzos who finds, as a result of horrendous discomfort experienced in the mind and body, that he or she is "not at liberty to stop," whether the situation comprises 'addiction' or 'dependency' is immaterial to the point of being ridiculous. What matters most to such a person is to find some way to rescue himself or herself from the predicament.

Chapter Twenty-Five
Getting Safely Off Benzos

Should you choose to discontinue your benzodiazepine use, a number of methods are available. The most common practice is probably to discuss the matter with your doctor, then to rely upon whatever procedure the doctor chooses. There is great variability in such procedures, however, and very little to suggest that they might be based upon methods derived from scientific or clinical evidence. Rather more likely is that what the doctor does will be based upon experience in tapering others, in which any symptoms of benzodiazepine withdrawal may have been overlooked—simply because the doctor never imagined they could be there. The doctor would then treat the symptoms in such a way as to obscure the actual problems, such treatment itself interfering with the course of discontinuation.

An alternative is to go to a detoxification facility to get off the drug, but accounts of the experiences of those who have undergone it suggest that detox is perhaps the second worst way to discontinue benzos, coming after 'cold turkey' abrupt cessation of the drug. The most dangerous aspect of discontinuation—potentially life threatening, in fact—is the risk of going into seizure from stopping too quickly. Detox centers almost invariably employ a carefully controlled program of dosing their patients with phenobarbital or other antiseizure drugs to keep them from going into convulsions while the benzodiazepine is rapidly reduced. The narcotic effect of these drugs spares patients from the immediate effects of withdrawal, keeping them relatively comfortable until the intake of benzodiazepine has been terminated. The process is effective in that patients may technically be rendered 'benzo free' at the end of it, but it does not address the true cause of benzodiazepine withdrawal syndrome. During the relatively short treatment period of from one to six weeks, patients discontinuing opiates, alcohol, and other substances of abuse often appear to recover their vitality, while those who were admitted for benzodiazepines languish in various states of malaise. Since the focus of such centers is upon substance *abuse,* not iatrogenic dependency, the psychological counseling they offer is centered on sobriety issues and other related topics which usually do not apply to benzodiazepine users. The same approach is also found in Twelve Step programs. While Alcoholics Anonymous and Narcotics Anonymous have helped many people, their emphasis on sobriety is inappropriate for most benzodiazepine users. Further, participants in such programs stress that members be 'drug and alcohol free' and may—with good intentions—exert their influence to get people quickly off benzos, though that would be inconsonant with what is known about successful discontinuation of them.

Most insurance policies will not pay for more than forty-five days of substance abuse care, so benzodiazepine discontinuation in detox facilities has to be tailored to the duration of treatment, rather than the other way around. People often leave fast-track detox facilities in a somewhat bewildered state, convinced they are now 'drug free' yet experiencing benzo withdrawal symptoms nonetheless. Because discontinuation can be a lengthy process, during which time the benzo user may often be far too debilitated to cope with the stresses of employment, the economic impact of large numbers of people removed from the work force and subsidized by the insurance industry while tapering off the drugs would be staggering. From this standpoint alone it is understandable that the medical establishment would continue its denial of the existence of a benzodiazepine withdrawal syndrome.

Fortunately, there exists the method developed by Prof. C. Heather Ashton during twelve years of operating a clinic devoted to benzodiazepine withdrawal. It is thoroughly detailed in her book, *Benzodiazepines: How They Work and How To Withdraw*, which can be purchased at *www.benzo.org.uk/bzmono.htm* for a $15 donation. It is also available on the Internet at no cost at *www.benzo.org.uk/manual/* where the book can be read online, downloaded, or printed from the computer. Anyone taking benzodiazepines would be well advised to read *Benzodiazepines: How They Work and How To Withdraw* as it contains all of the most relevant information, obtained by rigorous adherence to scientific principles and methods. Ray Nimmo's website, *www.benzo.org.uk,* in addition to the Ashton Manual, contains a vast amount of information relating to benzo withdrawal, everything from anecdotal accounts of the actual experiences people have had as a result of taking benzos, to news articles and numerous scientific papers. Understanding benzodiazepine withdrawal is perhaps the single most powerful asset anyone could possess should they decide to undergo the process; as such, information of the highest quality is vital.

Support, too, is most important. For those who have internet access, there are mailing lists and bulletin board groups available, such as:

www.benzoisland.org
www.benzobuddies.org
http://health.groups.yahoo.com/group/benzo

Not only do you get to hear other people's subjective opinion about what they're going through—which often may mirror and therefore validate your own experiences—there is also a great deal of help in everyday matters relating to reduction. For example, if you find that you have *cog fog* to contend with, others will share their own ways of coping with it (such as making lists of things not to forget and keeping it on the refrigerator) as well as how to deal with the feelings of inadequacy that having cognitive problems engenders. When peculiar or disturbing symptoms crop up, at least some others will have had them, too, and can reassure you that they are indeed benzo symptoms and will therefore eventually pass. And they will even remind you that there is sometimes a humorous side to having your world occasionally turned upside down. Never in my life had I been more needy of fellowship, and never more grateful than to have found it at *benzo.org.uk's* bulletin board.

Should you use Professor Ashton's method, you will require the compliance of a physician or psychiatrist in order to supply you with the necessary prescription for diazepam. It has been the experience of many that finding a doctor who will agree to take a participatory role in a benzodiazepine discontinuation plan using Valium can be difficult—even impossible for some. Oddly, and somewhat ironically,

Valium is perceived as a drug from the past that caused widespread addiction, while the newer benzos—far more potent and problematic—are thought to be safe. Therefore, some doctors refuse to write prescriptions for Valium. This is complicated by the fact that, due to the difference in potency, the amount of Valium it takes to equal your current dose of another benzodiazepine may well seem excessive to a doctor. A prescription for 4 mg of Klonopin daily, for example, would appear normal and perfectly acceptable, while prescribing 80 mg of Valium would seem like a massive amount of the drug, even though it is the equivalent of 4 mg of Klonopin.

Since having a supply of Valium is critical to a slow taper, it is best to make a convincing presentation of the request for a doctor's assistance. Rather than sitting in the doctor's office and asking for large quantities of Valium, a better method would be to deliver to the physician a copy of Professor Ashton's *Benzodiazepines: How They Work and How To Withdraw,* a copy of her paper, *Reasons for a diazepam (Valium) taper,* (available at *http://www.benzo.org.uk/ashvtaper.htm*), and a cover letter stating that you wish to go off the drug and that this is the safest, most documented method available. The letter should also say that if the doctor knows a better method, you would be happy for him to provide you with its documentation so you can compare the two. The doctor may actually have another tapering method—most probably quite rapid compared to Ashton's—but it is doubtful that it would be documented or supported by research. (APPENDIX H comprises just such a letter, which readers may copy and adapt to suit their own circumstances.) Also, it would be helpful if the information packet contains a printout of whatever tapering schedule (taken from *The Ashton Manual*) you intend to follow. The best way to present this would be to deliver it to the doctor well in advance of an appointment, saying that he or she should look over the material as you intend to discuss it in your next visit. This allows time for careful consideration beforehand to avoid the possibility that the doctor might get defensive should you present it during an office consultation, requiring an immediate decision which would result in prescribing what may seem like large amounts of Valium over a long period of time.

You will need to make a stepwise crossover from whatever benzodiazepine you are currently taking to diazepam. Because the action of each of the benzos is different, a sudden switch from any of them to diazepam could result in an insufficiency of benzodiazepine being available quickly enough, which the body would experience as a cold-turkey type withdrawal—even though you had ingested an equivalent amount diazepam tablets. *The Ashton Manual* provides complete details for accomplishing the crossover. Once it has been completed, and you are taking diazepam exclusively, you may begin to taper the drug.

One of the virtues of diazepam tablets that makes them convenient to use for tapering is that they come in 5 mg and 2 mg strengths, scored so they can be divided in half. Although they can be broken apart into two pieces with the fingers, a more precise division may be obtained by using a razor craft knife or an inexpensive pill cutter of the type sold at pharmacies. For doses that are multiples of five, the amount of pills to take is easy to determine: *e.g.*, a 30 mg dose consists of six 5 mg tablets. Whole numbered doses are also fairly easy to achieve. A dose of 29 mg, for example, would be made by combining five 5 mg tablets with two 2 mg tablets, and 28 mg by combining four 5 mg tablets with four 2 mg tablets. Later on in the taper, doses can be arrived at by cutting the tablets. A dose of 18.5 mg is comprised of three and one-half 5 mg tablets and one-half of a single 2 mg tablet. (See APPENDIX D for a table which shows the various combinations used to formulate dosage amounts.)

    Rather than fussing with the tablets every day, it is more efficient to make up all of a week's doses at one time. The focus this requires also helps to insure that mistakes aren't made. To this end, inexpensive pill organizers available at pharmacies and supermarkets are a great convenience. The simplest of them features seven connected boxes, each of which is marked with the day of the week. For those whose dose of diazepam is spread out throughout the day, there are pill organizers with additional compartments for each day, labeled *morning, noon, evening,* and *night.* Pill organizers offer the added advantage of helping keep track of when the tablets are taken, to avoid the risk of skipping a day's dose or taking too much diazepam on a given day. To keep track of the weekly dose changes, it is best to put a check mark on the tapering schedule as each stage is completed. It is also helpful to bring the tapering schedule to the physician or psychiatrist whenever it is time for the next prescription. For the doctor to have a visual corroboration that progress is being made will reinforce the idea that you are actually diminishing the benzodiazepine intake, providing an incentive to continue the program.

    These skills and procedures are all that is required for tapering with diazepam. To implement them usually does not require any self-discipline, unlike discontinuation from other addictive substances. The ever-decreasing amount of diazepam insures that there is enough of the drug available to keep the central nervous system from a state of hyperexcitability, diminishing the impact of the dose reductions. The schedules in the Ashton Manual suggest dosage reductions occur every week or two, but each person should determine the most comfortable rate for himself or herself. (Facsimiles of the schedules I adapted for my own use may be seen in APPENDIX B. Blank versions are available as APPENDIX E, APPENDIX F, and APPENDIX G, which follow the dosage tables in APPENDICES C and D.) During periods of extraordinary stress, or after a particularly rough cut, you may wish to stabilize at a given dosage for some extra time. It is not a good idea, however, to stay at one dose level for more than about three weeks. The experience of others suggests that after three weeks, *tolerance* may set in. It seems illogical to the point of being paradoxical that during tolerance, we tend to feel better from cutting than we would if we remained at a particular dose, but it is a widely reported phenomenon among diazepam taperers. So, even though withdrawal symptoms may escalate as the dosage comes down, the general sense of well-being appears to improve.

In the initial part of the taper, many people have reported feeling 'oversedated,' characterized by tiredness, slurred words, and a mild sense of intoxication; some have said that these feelings came as a relief. In any case, the phenomenon soon passes as the body acclimates to diazepam and the levels begin to drop as the first cuts are made. After that, the tapering rituals become routine, and the focus shifts to learning to tolerate such withdrawal symptoms as may arise and to getting on with the reductions. One of the biggest challenges is the length of time the process requires. And yet, the time does pass and 'the numbers,' *i.e.,* the size of the dose, continue to come down. As the dosage lowers, the size of the cuts is made smaller, which feels subjectively as though the process were being dragged out, but it is important to do this so as not to reduce by too great a percentage of the overall dose. After what may have seemed like a nearly interminable amount of time, the last dose is eventually taken. The taper is finally over.

Chapter Twenty-Six
Post-Benzo

As I have mentioned previously, some people begin to feel great as soon as there is no longer any benzodiazepine in their bodies, while others continue to experience a continuation of the same withdrawal symptoms they had during the tapering process. And others actually feel worse when they get off the drug. There is, unfortunately, no way to predict the outcome of discontinuation. According to Professor Ashton, writing in the 2004 edition of *Comprehensive Handbook of Drug & Alcohol Addiction*, 85 to 95% of those getting off benzos do not have related problems of any significance once the use of the drug has been terminated. The remaining 5 to 15%, however, may experience any of a number of withdrawal symptoms, which may include anxiety, insomnia, depression, cognitive impairment and gastrointestinal symptoms. She reports also the possibility of perceptual difficulties such as tinnitus, tingling, numbness or pain (usually in the limbs or extremities), and 'motor symptoms,' including muscle pain, weakness, tension, painful tremor, shaking attacks, jerks, and blepharospasm, an involuntary contraction of the eyelid. She reports that these components of a 'post withdrawal syndrome' usually gradually improve over the course of a year or more, although some may 'occasionally persist indefinitely.' Even so, there is little evidence to suggest that benzo use or withdrawal causes actual structural changes in the brain or central nervous system.

Unfortunately, the typical response by physicians to reports of bizarre, lingering symptoms after discontinuation is to believe that the patients making such complaints are experiencing the reemergence of previously existing problems which had been quelled by the use of benzodiazepines. Once the drug has been stopped, they reason, the conditions recur. When that idea is challenged by protests that the symptoms are nothing that the patients had ever had before taking the drugs, physicians evidently conclude that they have developed *new* neurological problems, whose pathogenesis is the neurotic nature of the patients themselves, not their having been exposed to benzodiazepine. The inability of these physicians to believe the testimony of their patients is a major contributory factor in benzo withdrawal having remained an unrecognized problem in so many people, over such a lengthy period of time.

Amongst the online population of people tapering with Valium, a very small percentage has reported a *Protracted Withdrawal Syndrome (PWS),* which has been defined as significant withdrawal symptoms continuing beyond eighteen months after benzodiazepine use has been terminated. Fortunately, this is apparently a somewhat rare occurrence. As Professor Ashton states in *The Ashton Manual,*
   All the evidence shows that a steady decline in symptoms almost

invariably continues after withdrawal, though it can take a long time—even several years in some cases. Most people experience a definite improvement over time so that symptoms gradually decrease to levels nowhere near as intense as in the early days of withdrawal, and eventually almost entirely disappear.

At the time of writing, I have been off benzodiazepine for nine months. I have seen gradual—*very* gradual—improvement in some areas, mostly in my cognitive abilities. Other withdrawal symptoms have persisted; they wax and wane in severity, as though my body can only sustain limited periods of up-regulated GABA utilization, then reverts to its previous state. Sleep has improved, yet the fatigue remains, although not nearly as severe as it had been. I do feel as though I am very slowly emerging from a long, strange dream, having inhabited for so long a time a grotesque nightmare world I could never have thought possible in my darkest imaginings. And I realize it could have been far worse had I not found Ray Nimmo's website and the research and methodology of Professor Ashton. I, and countless thousands of others, are indebted to them for providing a safe way out of the addiction doctors have unwittingly fostered in some of their patients.

While I am much improved compared to my condition during the years I spent tapering off of benzos—and vastly improved in comparison to the state I was in during the interdose withdrawal of the tolerance phase, I must recognize that I still live a diminished existence. All of my adult life, I felt capable of living in the world and doing something useful, something beneficial for others. Building things, doing things, solving problems, going places, working with people, such activities were so natural to me that I took them all for granted—until they were taken away. I felt deeply embarrassed by having a condition that minimized me to such a degree; it was especially painful for me that, in the eyes of my son, I had been reduced to someone who sat on the couch night after night, watching television, never saying very much or contributing much to our family. It was demeaning to think of myself as a victim—and yet, was not that term perfectly appropriate for someone who had been victimized by a medical system that refused to inform patients of the actual risks of its medicines?

The first glimmer of a possibility that I might once again participate in the life of the world came when I realized that enough of my mental abilities had been restored that I could write a book—this book. My second most important goal in writing it was to provide information that might conceivably help others who were facing the possibility of discontinuing benzodiazepine. The most important goal was to alert people to the methods of Prof. C. Heather Ashton, so they would have the best chance of coming off tranquilizers safely. Although still far from recovered, I am, thanks to Professor Ashton, free of benzos and once again full of that most necessary, most vital of human feelings—hope.

Appendices

APPENDIX A: Additional reading list

APPENDIX B: The author's tapering schedule sheets

APPENDIX C: Dosage equivalency table

APPENDIX D: Dosage quantity table

APPENDIX E: Alternate Schedule 2a

APPENDIX F: Alternate Schedule 2b

APPENDIX G: Alternate Shedule 3

APPENDIX H: Sample letter to doctor requesting use of the Ashton Method

Additional Reading

Prisoner On Prescription, Heather Jones
    Headway Books (September 30, 1990)
    ISBN: 0951304526

The Great Anxiety Escape, Max Ricketts
    Matulungin Publishing (June 1, 1990)
    ISBN: 0962620505

Your Drug May Be Your Problem, Peter Breggin, MD and David Cohen, PhD
    HarperCollins Publishers (August 1, 2000)
    ISBN: 0738203483

Benzo Blues, Edward Drummond, MD
    Plume Books (November 1, 1998)
    ISBN: 0452278260

Bitter Pills : Inside the Hazardous World of Legal Drugs, Stephen Fried
    Bantam (May 4, 1999)
    ISBN: 055337852X

Addiction by Prescription, Joan E. Gadsby
    Key Porter Books (March 1, 2000)
    ISBN: 1552633349

Swallowing a Bitter Pill: How Prescription and Over-The-Counter Drug Abuse Is Ruining Lives - My Story, Cindy R. Mogil
    New Horizon Press (November 1, 2001)
    ISBN: 088282211X

I'm Dancing as Fast as I Can, B. Gordon
    Harpercollins; Reissue edition (April 1989)
    ISBN: 0060915935

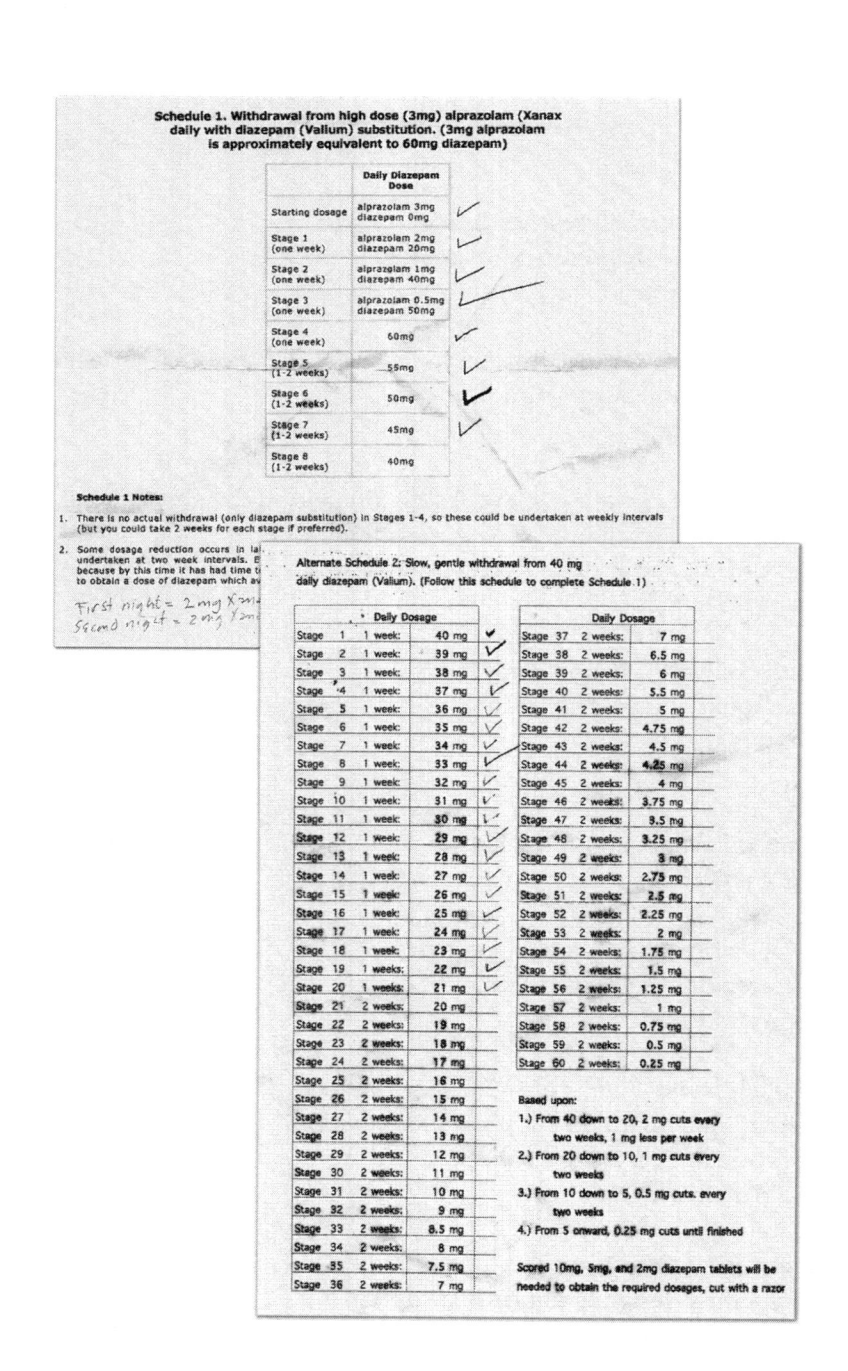

FIG. 1 : *The author's actual reduction schedules over the first twenty-eight weeks, each check mark representing one week*

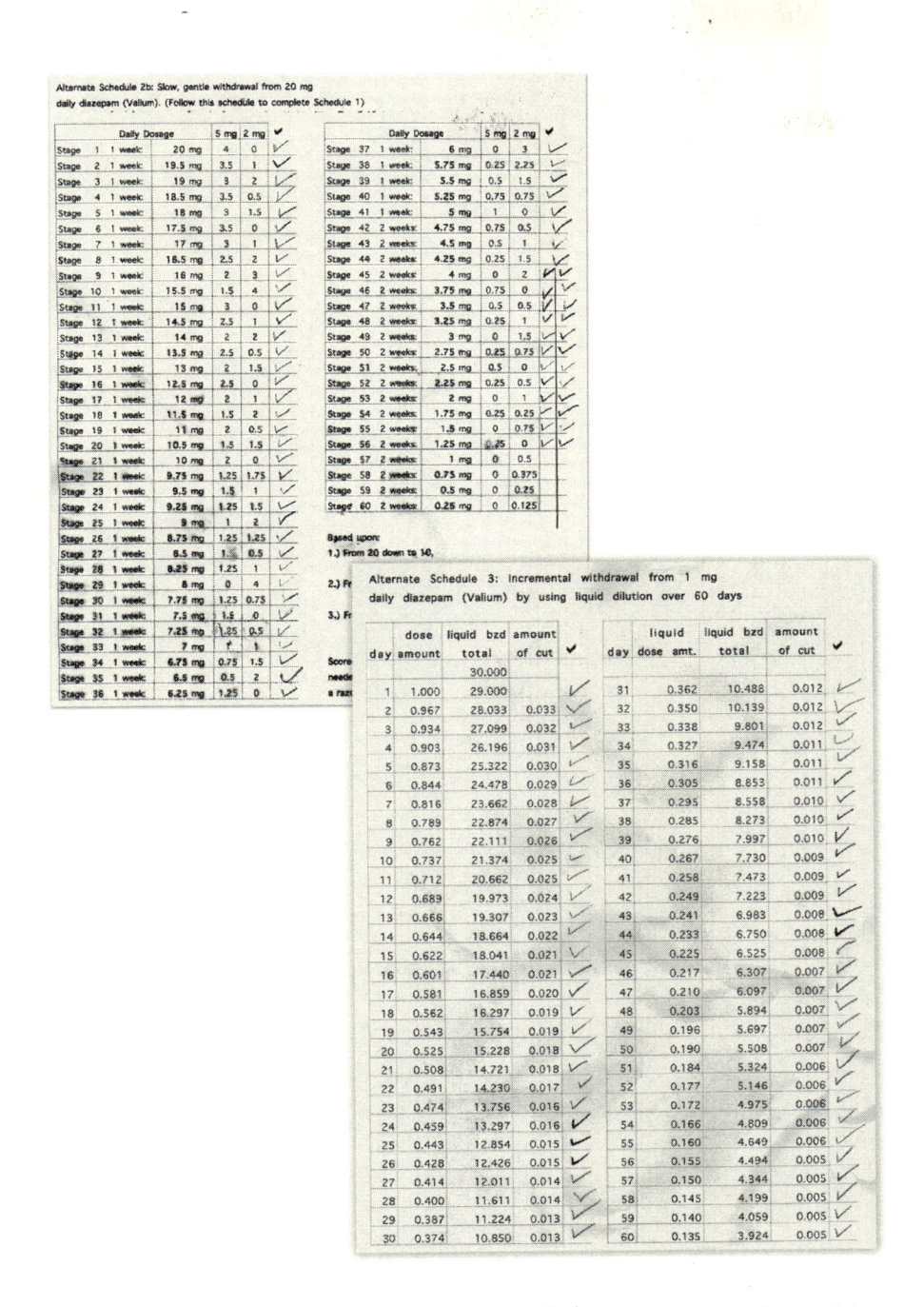

FIG. 2 : *The author's actual Schedule 2b (covering sixty-eight weeks) and Schedule 3 (covering the final eight weeks)*

THE BENZO BOOK

APPENDIX C

Dosage Equivalency

use this to derive any given dosage amount from 5 mg and 2mg tablets

Dosage	5 mg	2 mg	✔
20 mg	4	0	
19.5 mg	3.5	1	
19 mg	3	2	
18.5 mg	3.5	0.5	
18 mg	3	1.5	
17.5 mg	3.5	0	
17 mg	3	1	
16.5 mg	2.5	2	
16 mg	2	3	
15.5 mg	1.5	4	
15 mg	3	0	
14.5 mg	2.5	1	
14 mg	2	2	
13.5 mg	2.5	0.5	
13 mg	2	1.5	
12.5 mg	2.5	0	
12 mg	2	1	
11.5 mg	1.5	2	
11 mg	2	0.5	
10.5 mg	1.5	1.5	
10 mg	2	0	
9.75 mg	1.25	1.75	
9.5 mg	1.5	1	
9.25 mg	1.25	1.5	
9 mg	1	2	
8.75 mg	1.25	1.25	
8.5 mg	1.5	0.5	
8.25 mg	1.25	1	
8 mg	0	4	
7.75 mg	1.25	0.75	

Dosage	5 mg	2 mg	✔
7.5 mg	1.5	0	
7.25 mg	1.25	0.5	
7 mg	1	1	
6.75 mg	0.75	1.5	
6.5 mg	0.5	2	
6.25 mg	1.25	0	
6 mg	0	3	
5.75 mg	0.25	2.25	
5.5 mg	0.5	1.5	
5.25 mg	0.75	0.75	
5 mg	1	0	
4.75 mg	0.75	0.5	
4.5 mg	0.5	1	
4.25 mg	0.25	1.5	
4 mg	0	2	
3.75 mg	0.75	0	
3.5 mg	0.5	0.5	
3.25 mg	0.25	1	
3 mg	0	1.5	
2.75 mg	0.25	0.75	
2.5 mg	0.5	0	
2.25 mg	0.25	0.5	
2 mg	0	1	
1.75 mg	0.25	0.25	
1.5 mg	0	0.75	
1.25 mg	0.25	0	
1 mg	0	0.5	
0.75 mg	0	0.375	
0.5 mg	0	0.25	
0.25 mg	0	0.125	

© 2006 JACK HOBSON-DUPONT

Patient's Copy

THE BENZO BOOK　　　　　　　　　　　　　APPENDIX C

Dosage Equivalency

use this to derive any given dosage amount from 5 mg and 2mg tablets

Dosage	5 mg	2 mg	✔
20 mg	4	0	
19.5 mg	3.5	1	
19 mg	3	2	
18.5 mg	3.5	0.5	
18 mg	3	1.5	
17.5 mg	3.5	0	
17 mg	3	1	
16.5 mg	2.5	2	
16 mg	2	3	
15.5 mg	1.5	4	
15 mg	3	0	
14.5 mg	2.5	1	
14 mg	2	2	
13.5 mg	2.5	0.5	
13 mg	2	1.5	
12.5 mg	2.5	0	
12 mg	2	1	
11.5 mg	1.5	2	
11 mg	2	0.5	
10.5 mg	1.5	1.5	
10 mg	2	0	
9.75 mg	1.25	1.75	
9.5 mg	1.5	1	
9.25 mg	1.25	1.5	
9 mg	1	2	
8.75 mg	1.25	1.25	
8.5 mg	1.5	0.5	
8.25 mg	1.25	1	
8 mg	0	4	
7.75 mg	1.25	0.75	

Dosage	5 mg	2 mg	✔
7.5 mg	1.5	0	
7.25 mg	1.25	0.5	
7 mg	1	1	
6.75 mg	0.75	1.5	
6.5 mg	0.5	2	
6.25 mg	1.25	0	
6 mg	0	3	
5.75 mg	0.25	2.25	
5.5 mg	0.5	1.5	
5.25 mg	0.75	0.75	
5 mg	1	0	
4.75 mg	0.75	0.5	
4.5 mg	0.5	1	
4.25 mg	0.25	1.5	
4 mg	0	2	
3.75 mg	0.75	0	
3.5 mg	0.5	0.5	
3.25 mg	0.25	1	
3 mg	0	1.5	
2.75 mg	0.25	0.75	
2.5 mg	0.5	0	
2.25 mg	0.25	0.5	
2 mg	0	1	
1.75 mg	0.25	0.25	
1.5 mg	0	0.75	
1.25 mg	0.25	0	
1 mg	0	0.5	
0.75 mg	0	0.375	
0.5 mg	0	0.25	
0.25 mg	0	0.125	

© 2006 JACK HOBSON-DUPONT　　　　　　　　　Doctor's Copy

THE BENZO BOOK   APPENDIX D

Dosage Quantity

use this to determine how many of each tablet will need to be prescribed for a week's doses

Dosage	5 mg	2 mg	✔
20 mg	28	0	
19.5 mg	25	7	
19 mg	21	14	
18.5 mg	25	4	
18 mg	21	11	
17.5 mg	25	0	
17 mg	21	7	
16.5 mg	18	14	
16 mg	14	21	
15.5 mg	11	28	
15 mg	21	0	
14.5 mg	18	7	
14 mg	14	14	
13.5 mg	18	4	
13 mg	14	11	
12.5 mg	18	0	
12 mg	14	7	
11.5 mg	11	14	
11 mg	14	4	
10.5 mg	11	11	
10 mg	14	0	
9.75 mg	9	12	
9.5 mg	11	7	
9.25 mg	9	11	
9 mg	7	14	
8.75 mg	9	9	
8.5 mg	11	4	
8.25 mg	9	7	
8 mg	0	28	
7.75 mg	9	5	

Dosage	5 mg	2 mg	✔
7.5 mg	11	0	
7.25 mg	9	4	
7 mg	7	7	
6.75 mg	5	11	
6.5 mg	4	14	
6.25 mg	9	0	
6 mg	0	21	
5.75 mg	2	16	
5.5 mg	4	11	
5.25 mg	5	5	
5 mg	7	0	
4.75 mg	5	4	
4.5 mg	4	7	
4.25 mg	2	11	
4 mg	0	14	
3.75 mg	5	0	
3.5 mg	4	4	
3.25 mg	2	7	
3 mg	0	11	
2.75 mg	2	5	
2.5 mg	4	0	
2.25 mg	2	4	
2 mg	0	7	
1.75 mg	2	2	
1.5 mg	0	5	
1.25 mg	2	0	
1 mg	0	4	
0.75 mg	0	3	
0.5 mg	0	2	
0.25 mg	0	1	

© 2006 JACK HOBSON-DUPONT   Patient's Copy

THE BENZO BOOK　　　　　　　　　　　APPENDIX D

Dosage Quantity

use this to determine how many of each tablet will need to be prescribed for a week's doses

Dosage	5 mg	2 mg	✔
20 mg	28	0	
19.5 mg	25	7	
19 mg	21	14	
18.5 mg	25	4	
18 mg	21	11	
17.5 mg	25	0	
17 mg	21	7	
16.5 mg	18	14	
16 mg	14	21	
15.5 mg	11	28	
15 mg	21	0	
14.5 mg	18	7	
14 mg	14	14	
13.5 mg	18	4	
13 mg	14	11	
12.5 mg	18	0	
12 mg	14	7	
11.5 mg	11	14	
11 mg	14	4	
10.5 mg	11	11	
10 mg	14	0	
9.75 mg	9	12	
9.5 mg	11	7	
9.25 mg	9	11	
9 mg	7	14	
8.75 mg	9	9	
8.5 mg	11	4	
8.25 mg	9	7	
8 mg	0	28	
7.75 mg	9	5	

Dosage	5 mg	2 mg	✔
7.5 mg	11	0	
7.25 mg	9	4	
7 mg	7	7	
6.75 mg	5	11	
6.5 mg	4	14	
6.25 mg	9	0	
6 mg	0	21	
5.75 mg	2	16	
5.5 mg	4	11	
5.25 mg	5	5	
5 mg	7	0	
4.75 mg	5	4	
4.5 mg	4	7	
4.25 mg	2	11	
4 mg	0	14	
3.75 mg	5	0	
3.5 mg	4	4	
3.25 mg	2	7	
3 mg	0	11	
2.75 mg	2	5	
2.5 mg	4	0	
2.25 mg	2	4	
2 mg	0	7	
1.75 mg	2	2	
1.5 mg	0	5	
1.25 mg	2	0	
1 mg	0	4	
0.75 mg	0	3	
0.5 mg	0	2	
0.25 mg	0	1	

© 2006 JACK HOBSON-DUPONT

Doctor's Copy

Alternate Schedule 2a

slow, gentle withdrawal from 40 mg daily diazepam (Valium)

	Duration	Dosage	✔
1	1 week	40 mg	
2	1 week	39 mg	
3	1 week	38 mg	
4	1 week	37 mg	
5	1 week	36 mg	
6	1 week	35 mg	
7	1 week	34 mg	
8	1 week	33 mg	
9	1 week	32 mg	
10	1 week	31 mg	
11	1 week	30 mg	
12	1 week	29 mg	
13	1 week	28 mg	
14	1 week	27 mg	
15	1 week	26 mg	
16	1 week	25 mg	
17	1 week	24 mg	
18	1 week	23 mg	
19	1 week	22 mg	
20	1 week	21 mg	
21	2 weeks	20 mg	___
22	2 weeks	19 mg	___
23	2 weeks	18 mg	___
24	2 weeks	17 mg	___
25	2 weeks	16 mg	___
26	2 weeks	15 mg	___
27	2 weeks	14 mg	___
28	2 weeks	13 mg	___
29	2 weeks	12 mg	___
30	2 weeks	11 mg	___

	Duration	Dosage	✔
31	2 weeks	10 mg	___
32	2 weeks	9 mg	___
33	2 weeks	8.5 mg	___
34	2 weeks	8 mg	___
35	2 weeks	7.5 mg	___
36	2 weeks	7 mg	___
37	2 weeks	6.5 mg	___
38	2 weeks	6 mg	___
39	2 weeks	5.5 mg	___
40	2 weeks	5 mg	___
41	2 weeks	4.75 mg	___
42	2 weeks	4.5 mg	___
43	2 weeks	4.25 mg	___
44	2 weeks	4 mg	___
45	2 weeks	3.75 mg	___
46	2 weeks	3.5 mg	___
47	2 weeks	3.25 mg	___
48	2 weeks	3 mg	___
49	2 weeks	2.75 mg	___
50	2 weeks	2.5 mg	___
51	2 weeks	2.25 mg	___
52	2 weeks	2 mg	___
53	2 weeks	1.75 mg	___
54	2 weeks	1.5 mg	___
55	2 weeks	1.25 mg	___
56	2 weeks	1 mg	___
57	2 weeks	0.75 mg	___
58	2 weeks	0.5 mg	___
59	2 weeks	0.25 mg	___
60	2 weeks	0.125 mg	___

© 2006 JACK HOBSON-DUPONT

Patient's Copy

Alternate Schedule 2a
slow, gentle withdrawal from 40 mg daily diazepam (Valium)

	Duration	Dosage	✓
1	1 week	40 mg	
2	1 week	39 mg	
3	1 week	38 mg	
4	1 week	37 mg	
5	1 week	36 mg	
6	1 week	35 mg	
7	1 week	34 mg	
8	1 week	33 mg	
9	1 week	32 mg	
10	1 week	31 mg	
11	1 week	30 mg	
12	1 week	29 mg	
13	1 week	28 mg	
14	1 week	27 mg	
15	1 week	26 mg	
16	1 week	25 mg	
17	1 week	24 mg	
18	1 week	23 mg	
19	1 week	22 mg	
20	1 week	21 mg	
21	2 weeks	20 mg	
22	2 weeks	19 mg	
23	2 weeks	18 mg	
24	2 weeks	17 mg	
25	2 weeks	16 mg	
26	2 weeks	15 mg	
27	2 weeks	14 mg	
28	2 weeks	13 mg	
29	2 weeks	12 mg	
30	2 weeks	11 mg	

	Duration	Dosage	✓
31	2 weeks	10 mg	
32	2 weeks	9 mg	
33	2 weeks	8.5 mg	
34	2 weeks	8 mg	
35	2 weeks	7.5 mg	
36	2 weeks	7 mg	
37	2 weeks	6.5 mg	
38	2 weeks	6 mg	
39	2 weeks	5.5 mg	
40	2 weeks	5 mg	
41	2 weeks	4.75 mg	
42	2 weeks	4.5 mg	
43	2 weeks	4.25 mg	
44	2 weeks	4 mg	
45	2 weeks	3.75 mg	
46	2 weeks	3.5 mg	
47	2 weeks	3.25 mg	
48	2 weeks	3 mg	
49	2 weeks	2.75 mg	
50	2 weeks	2.5 mg	
51	2 weeks	2.25 mg	
52	2 weeks	2 mg	
53	2 weeks	1.75 mg	
54	2 weeks	1.5 mg	
55	2 weeks	1.25 mg	
56	2 weeks	1 mg	
57	2 weeks	0.75 mg	
58	2 weeks	0.5 mg	
59	2 weeks	0.25 mg	
60	2 weeks	0.125 mg	

© 2006 JACK HOBSON-DUPONT

Doctor's Copy

THE BENZO BOOK   APPENDIX F

Alternate Schedule 2b
expanded coverage of 20 mg down, showing 5 mg and 2 mg components

	Duration	Dosage	5 mg	2 mg	✔
21	1 week	20 mg	4	0	
22	1 week	19.5 mg	3.5	1	
23	1 week	19 mg	3	2	
24	1 week	18.5 mg	3.5	0.5	
25	1 week	18 mg	3	1.5	
26	1 week	17.5 mg	3.5	0	
27	1 week	17 mg	3	1	
28	1 week	16.5 mg	2.5	2	
29	1 week	16 mg	2	3	
30	1 week	15.5 mg	1.5	4	
31	1 week	15 mg	3	0	
32	1 week	14.5 mg	2.5	1	
33	1 week	14 mg	2	2	
34	1 week	13.5 mg	2.5	0.5	
35	1 week	13 mg	2	1.5	
36	1 week	12.5 mg	2.5	0	
37	1 week	12 mg	2	1	
38	1 week	11.5 mg	1.5	2	
39	1 week	11 mg	2	0.5	
40	1 week	10.5 mg	1.5	1.5	
41	1 week	10 mg	2	0	
42	1 week	9.75 mg	1.25	1.75	
43	1 week	9.5 mg	1.5	1	
44	1 week	9.25 mg	1.25	1.5	
45	1 week	9 mg	1	2	
46	1 week	8.75 mg	1.25	1.25	
47	1 week	8.5 mg	1.5	0.5	
48	1 week	8.25 mg	1.25	1	
49	1 week	8 mg	0	4	
50	1 week	7.75 mg	1.25	0.75	

	Duration	Dosage	5 mg	2 mg	✔
51	1 week	7.5 mg	1.5	0	
52	1 week	7.25 mg	1.25	0.5	
53	1 week	7 mg	1	1	
54	1 week	6.75 mg	0.75	1.5	
55	1 week	6.5 mg	0.5	2	
56	1 week	6.25 mg	1.25	0	
57	1 week	6 mg	0	3	
58	1 week	5.75 mg	0.25	2.25	
59	1 week	5.5 mg	0.5	1.5	
60	1 week	5.25 mg	0.75	0.75	
61	1 week	5 mg	1	0	
62	2 weeks	4.75 mg	0.75	0.5	
63	2 weeks	4.5 mg	0.5	1	
64	2 weeks	4.25 mg	0.25	1.5	
65	2 weeks	4 mg	0	2	
66	2 weeks	3.75 mg	0.75	0	
67	2 weeks	3.5 mg	0.5	0.5	
68	2 weeks	3.25 mg	0.25	1	
69	2 weeks	3 mg	0	1.5	
70	2 weeks	2.75 mg	0.25	0.75	
71	2 weeks	2.5 mg	0.5	0	
72	2 weeks	2.25 mg	0.25	0.5	
73	2 weeks	2 mg	0	1	
74	2 weeks	1.75 mg	0.25	0.25	
75	2 weeks	1.5 mg	0	0.75	
76	2 weeks	1.25 mg	0.25	0	
77	2 weeks	1 mg	0	0.5	
78	2 weeks	0.75 mg	0	0.375	
79	2 weeks	0.5 mg	0	0.25	
80	2 weeks	0.25 mg	0	0.125	

© 2006 JACK HOBSON-DUPONT

Patient's Copy

Alternate Schedule 2b
expanded coverage of 20 mg down, showing 5 mg and 2 mg components

	Duration	Dosage	5 mg	2 mg	✔
21	1 week	20 mg	4	0	
22	1 week	19.5 mg	3.5	1	
23	1 week	19 mg	3	2	
24	1 week	18.5 mg	3.5	0.5	
25	1 week	18 mg	3	1.5	
26	1 week	17.5 mg	3.5	0	
27	1 week	17 mg	3	1	
28	1 week	16.5 mg	2.5	2	
29	1 week	16 mg	2	3	
30	1 week	15.5 mg	1.5	4	
31	1 week	15 mg	3	0	
32	1 week	14.5 mg	2.5	1	
33	1 week	14 mg	2	2	
34	1 week	13.5 mg	2.5	0.5	
35	1 week	13 mg	2	1.5	
36	1 week	12.5 mg	2.5	0	
37	1 week	12 mg	2	1	
38	1 week	11.5 mg	1.5	2	
39	1 week	11 mg	2	0.5	
40	1 week	10.5 mg	1.5	1.5	
41	1 week	10 mg	2	0	
42	1 week	9.75 mg	1.25	1.75	
43	1 week	9.5 mg	1.5	1	
44	1 week	9.25 mg	1.25	1.5	
45	1 week	9 mg	1	2	
46	1 week	8.75 mg	1.25	1.25	
47	1 week	8.5 mg	1.5	0.5	
48	1 week	8.25 mg	1.25	1	
49	1 week	8 mg	0	4	
50	1 week	7.75 mg	1.25	0.75	

	Duration	Dosage	5 mg	2 mg	✔
51	1 week	7.5 mg	1.5	0	
52	1 week	7.25 mg	1.25	0.5	
53	1 week	7 mg	1	1	
54	1 week	6.75 mg	0.75	1.5	
55	1 week	6.5 mg	0.5	2	
56	1 week	6.25 mg	1.25	0	
57	1 week	6 mg	0	3	
58	1 week	5.75 mg	0.25	2.25	
59	1 week	5.5 mg	0.5	1.5	
60	1 week	5.25 mg	0.75	0.75	
61	1 week	5 mg	1	0	
62	2 weeks	4.75 mg	0.75	0.5	
63	2 weeks	4.5 mg	0.5	1	
64	2 weeks	4.25 mg	0.25	1.5	
65	2 weeks	4 mg	0	2	
66	2 weeks	3.75 mg	0.75	0	
67	2 weeks	3.5 mg	0.5	0.5	
68	2 weeks	3.25 mg	0.25	1	
69	2 weeks	3 mg	0	1.5	
70	2 weeks	2.75 mg	0.25	0.75	
71	2 weeks	2.5 mg	0.5	0	
72	2 weeks	2.25 mg	0.25	0.5	
73	2 weeks	2 mg	0	1	
74	2 weeks	1.75 mg	0.25	0.25	
75	2 weeks	1.5 mg	0	0.75	
76	2 weeks	1.25 mg	0.25	0	
77	2 weeks	1 mg	0	0.5	
78	2 weeks	0.75 mg	0	0.375	
79	2 weeks	0.5 mg	0	0.25	
80	2 weeks	0.25 mg	0	0.125	

© 2006 JACK HOBSON-DUPONT

Doctor's Copy

Alternate Schedule 3

incremental withdrawal from 1 mg daily diazepam (Valium) using liquid dilution over 60 days

day	dose	liquid bzd	amount	✔
1	1.000	29.000		
2	0.967	28.033	0.033	
3	0.934	27.099	0.032	
4	0.903	26.196	0.031	
5	0.873	25.322	0.030	
6	0.844	24.478	0.029	
7	0.816	23.662	0.028	
8	0.789	22.874	0.027	
9	0.762	22.111	0.026	
10	0.737	21.374	0.025	
11	0.712	20.662	0.025	
12	0.689	19.973	0.024	
13	0.666	19.307	0.023	
14	0.644	18.664	0.022	
15	0.622	18.401	0.021	
16	0.601	17.440	0.021	
17	0.581	16.859	0.020	
18	0.562	16.297	0.019	
19	0.543	15.754	0.019	
20	0.525	15.228	0.018	
21	0.508	14.721	0.018	
22	0.491	14.230	0.017	
23	0.474	13.756	0.016	
24	0.459	13.297	0.016	
25	0.443	12.854	0.015	
26	0.428	12.426	0.015	
27	0.414	12.011	0.014	
28	0.400	11.611	0.014	
29	0.387	11.224	0.013	
30	0.375	10.850	0.013	

day	dose	liquid bzd	amount	✔
31	0.362	10.488	0.012	
32	0.350	10.139	0.012	
33	0.338	9.801	0.012	
34	0.327	9.474	0.011	
35	0.316	9.158	0.011	
36	0.305	8.853	0.011	
37	0.295	8.558	0.010	
38	0.285	8.273	0.010	
39	0.276	7.997	0.010	
40	0.267	7.730	0.009	
41	0.258	7.473	0.009	
42	0.249	7.223	0.009	
43	0.241	6.983	0.008	
44	0.233	6.750	0.008	
45	0.225	6.525	0.008	
46	0.217	6.307	0.007	
47	0.210	6.097	0.007	
48	0.203	5.894	0.007	
49	0.196	5.697	0.007	
50	0.190	5.508	0.007	
51	0.184	5.324	0.006	
52	0.177	5.146	0.006	
53	0.172	4.975	0.006	
54	0.166	4.809	0.006	
55	0.160	4.649	0.006	
56	0.155	4.494	0.005	
57	0.150	4.344	0.005	
58	0.145	4.199	0.005	
59	0.140	4.059	0.005	
60	0.135	3.924	0.005	

© 2006 JACK HOBSON-DUPONT

Patient's Copy

Alternate Schedule 3

incremental withdrawal from 1 mg daily diazepam (Valium) using liquid dilution over 60 days

day	dose	liquid bzd	amount
1	1.000	29.000	
2	0.967	28.033	0.033
3	0.934	27.099	0.032
4	0.903	26.196	0.031
5	0.873	25.322	0.030
6	0.844	24.478	0.029
7	0.816	23.662	0.028
8	0.789	22.874	0.027
9	0.762	22.111	0.026
10	0.737	21.374	0.025
11	0.712	20.662	0.025
12	0.689	19.973	0.024
13	0.666	19.307	0.023
14	0.644	18.664	0.022
15	0.622	18.041	0.021
16	0.601	17.440	0.021
17	0.581	16.859	0.020
18	0.562	16.297	0.019
19	0.543	15.754	0.019
20	0.525	15.228	0.018
21	0.508	14.721	0.018
22	0.491	14.230	0.017
23	0.474	13.756	0.016
24	0.459	13.297	0.016
25	0.443	12.854	0.015
26	0.428	12.426	0.015
27	0.414	12.011	0.014
28	0.400	11.611	0.014
29	0.387	11.224	0.013
30	0.375	10.850	0.013

day	dose	liquid bzd	amount
31	0.362	10.488	0.012
32	0.350	10.139	0.012
33	0.338	9.801	0.012
34	0.327	9.474	0.011
35	0.316	9.158	0.011
36	0.305	8.853	0.011
37	0.295	8.558	0.010
38	0.285	8.273	0.010
39	0.276	7.997	0.010
40	0.267	7.730	0.009
41	0.258	7.473	0.009
42	0.249	7.223	0.009
43	0.241	6.983	0.008
44	0.233	6.750	0.008
45	0.225	6.525	0.008
46	0.217	6.307	0.007
47	0.210	6.097	0.007
48	0.203	5.894	0.007
49	0.196	5.697	0.007
50	0.190	5.508	0.007
51	0.184	5.324	0.006
52	0.177	5.146	0.006
53	0.172	4.975	0.006
54	0.166	4.809	0.006
55	0.160	4.649	0.006
56	0.155	4.494	0.005
57	0.150	4.344	0.005
58	0.145	4.199	0.005
59	0.140	4.059	0.005
60	0.135	3.924	0.005

© 2006 JACK HOBSON-DUPONT

Doctor's Copy

Professor C. Heather Ashton's book, *Benzodiazepines: How They Work and How To Withdraw,* also known as *The Ashton Manual,* is available at:
http://www.benzo.org.uk/manual/index.htm
.... where it may be printed out directly from the computer. Hard copies of the book may be purchased at:
http://www.benzo.org.uk/bzmono.htm#order

Her paper entitled, *"Reasons for a Diazepam (Valium) Taper",* may be found at:
http://www.benzo.org.uk/ashvtaper.htm

It is strongly advised that copies of these two documents should be presented to the prescribing physician along with the accompanying letter. When keeping the initial appointment with the doctor to begin the discontinuation process, it is further advisable to bring copies of the tapering schedules, either from *The Ashton Manual* or those found on preceding pages of this Appendix. Doctors may tend to view information in printed form as more authoritative than if it is conveyed merely verbally.

Dear Dr. [name of doctor],

I have decided that I would like to discontinue my use of [name of drug] and will schedule an appointment with you to begin the process. Since [name of drug] is a benzodiazepine drug, I understand that I will need to taper off of it over a period of time. I have looked into this subject extensively and have determined that the tapering method best supported by both clinical studies and practical results is the one developed by Prof. C.H. Ashton, Emeritus Professor of Clinical Psychopharmacology at the University of Newcastle in England. The techniques of her method are presented in her book, "Benzodiazepines: How They Work and How To Withdraw". I have included a copy of the book for you to review before my appointment with you, should you feel it necessary.

Because of its longer half-life, diazepam is the benzodiazepine used for the tapering period. Prof. Ashton's rationale for this is summarized in her paper, "Reasons for a Diazepam (Valium) Taper", a copy of which accompanies this letter.

If you have an alternative method, I will be glad to look over any peer-reviewed papers or studies, or other material you have that pertains to it.

Thank you,
[your name]

Index

Acetlycholine, 83
Addiction, 7, 11, 14, 28, 35–37, 41, 80, 98, 100, 119, 121, 144, 167, 169, 189, 193–197, 202, 208
Adler, Meir, 69
Adrenaline, 45, 96
Adverse Psychiatric Reactions Information Link, 27
Adverse Reaction Reports, 55–56, 58
Agoraphobia, 19, 78–79, 127, 175, 193
Alcohol, 17, 38, 51, 80, 82, 84, 89, 175–176, 190, 199, 206
Alcoholics Anonymous, 199
Alexander, Dr. Bruce, 172
Alliance for Human Research Protection, 113
Alprazolam (Xanax), 6–14, 16, 18, 21, 23–24, 29, 33–35, 37, 44–45, 47, 52, 72–73, 98–99, 113, 125, 128, 140–141, 145, 173–178, 180, 189, 191, 196
American Academy of Family Physicians, 40, 189
Amnesia, 56, 58, 72
Anger, 14, 94
Anhedonia, 84, 130–131, 142, 168
Annual Death Rate, 69
Anorgasmia, 128
Antibiotics, 33, 50, 67, 180, 182–185
Antidepressant, 5, 7, 11, 106, 109, 111–112, 114–115, 119, 159, 174–175
Antipsychotic Drugs, 107, 123
Antiviral Drugs Advisory Committee, 63
Anxiety, 19, 26–27, 29–30, 40–41, 45–46, 51, 56, 75, 77–79, 82–84, 87–89, 90, 92–96, 98, 101, 103, 104, 159–161, 172, 178–179, 181, 184, 186–191, 193–194, 196, 206
Argentina, 59
Ashton, Prof. C. Heather, 9–10, 12–14, 39–40, 90–91, 100, 136, 139, 152–153, 179, 184, 200–202, 206–208
Autonomic Nervous System, 80

B Vitamins, 89
Babesiosis, 4–5, 180
Barbiturates, 40, 82
Barnes, Dr. Connie L., 182
Benzheptoxdiazine, 50
benzo.org.uk, 16, 18,–19, 35–36, 76, 79, 99, 127, 133, 140, 146, 152, 164, 167, 180, 200–202
benzobuddies.org, 201
Benzodiazepine, 9–16, 18–20, 25–31, 33–42, 44, 49, 50–52, 54, 56, 71–78, 80, 82–91, 94, 95, 96, 97, 101, 102, 103–104, 107, 113, 115, 124–128, 130–131, 133, 136–137, 139–141, 144–148, 150–151, 153–154, 157, 163–165, 167–181, 184–204, 206–208
Benzodiazepines: How They Work And How To Withdraw, 9–10, 13, 167, 179, 202–203, 207
benzoisland.org, 201
Biagini, Marco Biagini, 182
Biodefense and Pandemic Vaccine and Drug Development Act of 2005, 60
Biomedical Advanced Research and Development Agency, 60
Brackbill, Dr. Marcia L., 182
Brazil, 59
Breggin, Dr. Peter R., 113
British National Formulary, 26, 30
Buproprion, 111
Burroughs, Dr. Richard, 62

Controlled Substances, 51, 181
Calcium, 88
Cauchon, Dennis, 65
Center for Drug Evaluation and Research, 55
Center for Responsible Politics, 61
Central Nervous System, 6, 80, 83, 101, 104, 107, 115, 139, 166, 180–181, 184, 204, 207
Chemical Imbalance, 106, 108–109, 111–113, 114
Chronic Fatigue Syndrome, 5, 127, 168
Ciprofloxacin, 181
Citizens for Better Medicare, 59
Clinical Trial, 53, 55, 57, 113
Clonazepam (Klonopin), 52, 174–175, 178, 187, 189, 191, 193–194, 202
Cocaine, 38, 98
Cog Fog, 131–132, 201
Cognitive Behavior Therapy, 45
Cold Turkey, 10, 36, 199

Colombia, 69
Committee on The Review of Medicines, 27
Compassion Burnout, 144, 146
Conservative Mode, 129–130
Cooperstock, Dr. Ruth, 51
Cranial Electrotherapy Stimulation, 159–163
Cullen, Alicia, 181

Dalmane, 9, 51, 56, 58, 178
Dean, Dr. Carolyn, 68
Denmark, 59
Dependency, 7, 9, 35–36, 41, 69, 124, 177, 187, 190, 192, 194–195, 197, 199
Depersonalization, 29, 133, 140
Depression, 5, 19, 58, 78–79, 87–88, 101–102, 104, 105–109, 111–113, 115, 118, 142, 156–161, 164, 179–180, 187, 206
Derealization, 133, 140
Detoxification, 7, 36, 172, 174–176, 199, 200
Diabetes, 110, 124
Diazepam (Valium), 9–19, 21, 23, 34–35, 37, 45, 50–52, 71–72, 75, 88, 90, 92, 98–99, 113, 125, 127–128, 145, 148–149, 152, 164–167, 175–176, 191, 196, 201–205, 207
Discontinuation, 13, 40–41, 44, 77–79, 83–85, 87–89, 90, 95, 100, 102, 107, 124, 126, 131, 136–137, 139, 143, 144–145, 148, 151, 153, 159, 172–174, 189, 191–192, 198–201, 204, 206
Dosage, 8, 9, 12, 13, 29, 34, 35, 36, 37, 52, 71, 72, 74, 75, 87, 91, 98, 99, 108, 114, 125, 149, 152, 153, 154, 164, 165, 166, 189, 190, 191, 194, 195, 196, 204, 205
Down-regulation, 39, 40, 41, 74, 83, 84, 100, 128, 130, 142, 145, 157, 163, 167, 174, 176, 185, 193
Dreams, 44, 45, 208
Drug Addiction, 163, 195
Dysphoria, 84, 103, 142, 172

Edronax, 111
Elashoff, Dr. Michael, PhD, 62
Eli Lilly & Company, 113, 115
Emotional Blunting, 136–137, 139, 150, 179
Epstein-Barr Virus, 5
Equivalency Table, 10
Excitatory State, 40, 80, 89–90, 94, 142, 184

Fasciculations, 75–78, 83, 166
Fatigue, 3, 5, 44–46, 58, 74, 115, 126, 134, 144, 160–161, 168, 180, 207

Fear, 18–19, 45, 93–95, 103, 155–156, 158, 164, 171
Feldman, Dr. Martin, 68
Fetto, John, 116
Fight/Flight Response, 94
Flunitzepam (Rohypnol), 72
Fluoroquinolone Antibiotics, 180–185
Fortune 500, 59
France, 55
Freedom of Information Act, 54, 60
Frist, Senator Bill, 62
Fuerst, Dr. Samuel, 53

GABA, 38–42, 74, 77, 80–91, 94–95, 99, 103, 106, 128, 130, 139–142, 146, 149, 151, 154, 157, 163–164, 167, 170, 174, 176–177, 184–185, 193, 207
Gabapentin (Neurontin), 87, 124, 187
Games, 133–136
Gatifloxacin, 181
Generalized Anxiety Disorder, 172, 187, 190, 192–193
Glaxo Wellcome Pharmaceuticals, 63
GlaxoSmithKline, 63, 114
Glenmullen, Dr. Joseph, 110
Grenard, Steve, 44

H. Pylori, 32
Half-Life, 9, 12, 16, 34, 72, 152
Hamilton, Dr. John D., 63
Harrell, Richard M., 182
healthyplace.com, 190
Healy, Dr. David, 197
Heroin, 10, 38, 98
Holland, 52–53, 55

Iatrogenic Addiction, 121
Informed Consent, 11, 70, 116, 124, 171
Insomnia, 26–28, 40–42, 54, 58, 77–79, 82–84, 107, 123–124, 126, 128, 159–161, 175, 178, 181, 184, 189, 196, 206
Institute of Medicine, 57–58
Interdose Withdrawal, 9, 34, 128
Ireland, 114
Irish Medicines Board, 114

Johns Hopkins School of Hygiene and Public Health, 68
Johnson & Johnson Pharmaceuticals, 184
Johnson, Dr. Brian, 28, 40

Kennedy, Sen. Edward, 51
Kilkenny, Thomas M., 44
Koranyi, Dr. Erwin, 110
Kravitz, Dr. Richard L., 109

Lacasse, Jeffrey R., 111–112
Leo, Jonathan, 111
Lethargy, 46, 78–79, 178
Levaquin, 180–181, 184
Levitra, 115
Levofloxacin, 181–182
Libido, 79–80, 115, 127–128, 130–131, 156, 174
Librium, 49–50, 52, 191
Ligand-gated Neurons, 81
Light Therapy, 158
Liquid Dilution Method, 164, 167
Longo, Dr. Lance P., 28, 40
Lorazepam (Ativan), 52, 113, 175–176, 180, 191, 193
Lund, Dr. Brian C., 172
Lust, 93, 95, 130

Magnesium, 87, 88
Malaguti, Moreno, 182
Marshall, Dr. Barry, 32
Memory, 58, 72–74, 78, 79, 133–134, 179
Menzies, Karen Barth, 112
Merck Pharmaceuticals, 61
Methylamine, 50
Miltown, 49–50
Muscle Spasms, 58, 78, 82, 83

Narcotics Anonymous, 199
National Panic and Anxiety Disorder News, 188
Neural Pathways, 80
Neural Receptors, 38, 71, 85, 139
Neurotransmitters, 80–81, 94, 106, 110, 128
Niacin, 86, 88

Niacinamide, 86, 88
Nicotine, 38
Nimmo, Ray, 16, 18–19, 76, 79, 146–147, 200, 208
Norepinephrine, 106, 111
Norway, 59
Null, Gary, PhD, 68
Nutriceuticals, 87

Omniflox, 181
Opiates, 80, 82, 199
Ofloxacin, 181

Panic, 78, 103, 156, 178, 188, 190–191, 193
Parasympathetic Nervous System, 83, 84
Passionflower, 88
Pattison, Neal, 59
Perry, Dr. Paul, 172
Pharmaceutical Companies, 48, 59–62, 64–67, 86, 109, 112, 116–118, 120, 122, 124, 163
Picamilon, 86, 88
Pregabalin (Lyrica), 87
Productive Mode, 129
Project Bioshield Act of 2003, 60
PROTOCOL 321, 53–54
PROTOCOL 6415, 53
Protracted Withdrawal Syndrome, 90, 207
Provigil, 115
Prozac, 7, 98, 108–110, 113, 115, 118
Psychopharmacological Drugs Advisory Committee, 56
Psychotropic Drugs, 13
Public Citizen, 59
Pfizer Pharmaceuticals, 87

Quigley, Paul A., 25

Rage, 79, 93–95
Rasio, Dr. Debora, 68
Rebound, 44, 83–84, 137, 172, 188–189, 191
Reboxetine, 111
Relenza, 63–64, 67
REM Sleep, 44

Restless Leg Syndrome, 75, 166
Restoril, 9, 51, 56, 58, 178
Risk:Benefit Ratio, 123
RO 5-0690, 50
Roche Pharmaceuticals, 49, 50, 173
Rohypnol, 9, 72

Sam-E, 88
Seizure, 199
Selective Serotonin Reuptake Inhibitor, 5, 7, 107, 111–114, 118–119, 136–137, 157–158, 175
Self-Talk, 96, 141
Serotonin, 84, 106, 110–111, 113, 157–158
Sex Drive, 128
Sexual Dysfunction, 128, 137
Shaw, Dr. Robert, 53
Side Effects, 8, 28, 50, 53, 55–56, 58, 69, 87, 98, 114, 122, 158–162, 174, 177, 178, 180–181, 183–184, 190, 193
Sleep, 2–4, 6–7, 15, 42–44, 56, 58, 74, 80, 83, 88, 93, 106, 115, 125–126, 128, 157, 163, 168, 175, 178
Smith, Dr. Dorothy, PhD, 68
Social Anxiety Disorder, 109, 178, 193
Social Phobia, 187, 192
socialanxiety.factsforhealth.org, 187, 192
socialfear.com, 187, 191
Soviet Union, 159
Sternbach, Dr. Leo, 49–50
Substance Abuse, 7, 175, 189, 191–192, 193–196, 199–200
Suicidal Ideation, 19, 79–80, 141, 156, 158, 164

Tapering, 13, 18, 27, 71, 73, 88, 90, 98, 115, 127–128, 145, 151–154, 164, 167, 168, 188–189, 192, 196, 198, 200, 202–207
Tapering Schedule, 13, 152, 153, 202, 204
Tardive Dyskinesia, 124
The Pew Internet & American Life Project, 186
Theanine, 86, 88
toadgames.com, 133–134, 136
Tolerance, 9, 29, 40, 45, 137, 145, 172, 177, 179–180, 185, 190, 195–197, 205
Tranquilizer, 7, 49–50, 73, 77–78, 98, 137
TRANX, 25, 78
Triazolam (Halcion), 9, 52–58, 67

Triolo, Luigi, 182
Trovafloxacin, 181

U.S. Chamber of Commerce, 59
U.S. Food and Drug Administration, 26, 30, 49, 52–57, 59, 62–67, 99, 109, 112–117, 120, 162–163, 183–184, 192
Uhlenhuth, Dr. E.H., 29
United Kingdom, 10, 19, 25–26, 30, 51, 55, 59, 147
United States, 2, 10, 24, 26, 30, 49, 51–52, 59, 66, 68–69, 109, 115–116, 159, 163
University of Maryland Medical Center, 190
Upjohn, 29, 52–56

Valerian, 88
Venditiogenic Addiction, 121–122
Vestra, 111
Viagra, 115
Vitamin B, 89

Wallace Pharmaceuticals, 49
Warren, Dr. Robin, 32
Warren, Luke, 59
Wellbutrin, 111
Windows, 2, 42, 150–151, 157, 169
Withdrawal, 9, 13–14, 25, 28–30, 36–37, 40–41, 45, 73–80, 82–84, 87–89, 92, 94, 96–97, 103, 127–128, 130, 132–134, 136, 138, 140–144, 146–147, 150–152, 155, 157, 163–165, 167–181, 184–185, 188–189, 191, 193–200, 203, 205–207
Withdrawal Symptoms, 25, 30, 35, 37, 40–41, 74, 75–80, 82, 84, 87, 96, 152, 164, 165–168, 172–174, 180, 184, 188–189, 191, 194, 196, 200, 205–207
Woo, Dr. Theresa, 52
Word Jumble, 135
Word Noodle, 134
Work, David R., 181

Zocor, 61

Printed in Great Britain
by Amazon.co.uk, Ltd.,
Marston Gate.